QUANTUM PHYSICS FOR HIPPIES

Lukas Neumeier

James Douglas

AUTHORS

Dr. Lukas Neumeier

Born in Germany (1985). He received his Ph.D. in quantum nano-photonics from ICFO, Barcelona in 2018. He is still doing research in quantum physics at the University of Vienna.

Dr. James Douglas

Born in New Zealand and living in Barcelona. He received his D.Phil. from Oxford University in 2011 for work on ultracold quantum gases.

Jun Matsuura

Born in Barcelona (1980) with Japanese roots. He has been a professional drawing artist since 2004, focusing mainly on pencil drawing techniques.

Quantum Physics for Hippies Copyright © 2019 by Dr. Lukas Neumeier. All Rights Reserved.

No part of this book may be reproduced in any form or by any electronic or mechanical means including information storage and retrieval systems, without permission in writing from the author. The only exception is by a reviewer, who may quote short excerpts in a review.

Cover designed by Jun Matsuura

This book is a work of fiction. Names, characters, places, and incidents either are products of the author's imagination or are used fictitiously. Any resemblance to actual persons, living or dead, events, or locales is entirely coincidental.

Lukas Neumeier
Visit my website at www.hippie-nerd.com

First Printing: April 2019

Amazon

ISBN: 9781091891166

For all hippies, all nerds, everyone who is neither and everyone who is both.

PROLOGUE

Somewhere in Barcelona.
"You're a quantum physicist? Fascinating! I love quantum physics! It's about how we are all cosmic vibrations resonating with the universe, right?"

"No, it's not."

"But I read that in a book!"

"Sorry to be so blunt, but that book you read is bullshit."

"Hmm okay. So what is quantum physics about then?"

"Quantum physics is about everything. Let's grab a beer and talk about it if you like."

You might not believe it, but we have countless conversations like this. Most of the time they are fun. But sometimes we just want to dance, relax with friends or stand alone in the corner fulfilling our duty as introverted nerds. So to avoid feeling bad because people were missing out on the true magic of quantum physics, we decided to write a book. A book explaining what quantum physics is really about in a way a ten-year-old could understand. Or a grown-up after a few drinks. A book based on facts that still blows your mind.

Why did we call it *Quantum Physics for Hippies*? One reason is that hippies are a critically endangered species when it comes to misinformation about quantum physics. But more importantly, we chose the name because hippies and nerds have something unexpected in common. While nerds read books about physics, cosmology and computer science, hippies go to meditation retreats and emotional healing workshops. Do you see the connection?

Even though these things happen in different realms, in the end, each hippie and each nerd is simply looking for truth. The truth about who we are and how we fit in the universe. This search unites us all, regardless of social conventions.

How does quantum physics, a strange theory that describes the behaviour of atoms and light, help us in this search? It helps by telling us that the rules underpinning every aspect of reality are not those we expect. Besides revealing severe mistakes in the way we perceive the physical world, it also questions fundamental beliefs we have about ourselves. Why? Because we are made of atoms too.

This book follows a conversation between a nerd and a hippie as they chat about the mysteries of quantum physics, beginning with the dance of possibilities, Schrödinger's cat, quantum entanglement, and the influence of gossip. The conversation then moves on to what that all means for us as they discuss the possibility wave of the universe, parallel universes, consciousness, the feeling of identity and the question of who or what we really are. This leads to a surprise that will shake your sense of reality completely.

We hope you enjoy the journey down the rabbit hole of quantum reality as much as we do. Don't worry if you don't understand everything. Nobody does.

QUANTUM PHYSICS FOR HIPPIES

> *"The most beautiful experience we can have is the mysterious. It is the fundamental emotion that stands at the cradle of true art and true science."*
> -Albert Einstein

As I walked down Carrer de Paris in Barcelona, I had no idea I was heading for an audience with death.

What a beautiful day. I'm curiously looking into the windows of the cafes and restaurants as I am passing by. No plan at all and enough money to survive for almost a month. I just arrived from a week of yoga and meditation in the Pyrenees.

Hosting a couple of tantra sessions brought me a few bucks. I also attended some healing workshops. I'm trying to resolve my destructive conditionings, and man, do I have a few.

They say that the negative beliefs we have about ourselves are physically encoded in our brains[1]. Like little circuits. If you want to get rid of them, you can't simply acquire new positive beliefs. New beliefs just compete with the old ones, draining your energy. And sooner or later, life is going to find a way of triggering those old circuits again. The only smart thing you can do is to directly erase them. Aka shadow work. Shadow, because the old circuits are hidden behind a curtain. You feel the emotions they trigger, but you don't know why. You feel anger or fear and end up sabotaging your relationships because you blame your partner for those feelings. To me, shadow work is the most effective thing a human can do to improve their life. And tantra, by using intimacy to trigger our deepest beliefs about ourselves and our relationships, is a way to expose and overwrite those old circuits. This is why I do it. And because of the sex.

The Pyrenees is a beautiful place to escape to. Every time I go, I come back to the city refreshed. But this time I'm left mulling over one particular incident. It was this tantra ses-

---

[1] Ecker, Bruce. "Clinical translation of memory reconsolidation research: Therapeutic methodology for transformational change by erasing implicit emotional learnings driving symptom production." (2017).

sion with Samantha, a quantum healer. I must admit, it confuses me a lot. Even though we didn't touch each other, she came. Just by looking into my eyes. After she woke up from her blackout, she gave me her version of what happened. "We entangled each other," she said. "Do you know about quantum physics and Schrödinger's cat?"

I did not. Why the hell should I care about the cat of someone I had never heard of? Well, I really should have listened.

I walk slowly. I love to walk slowly. I don't understand people who seem to run everywhere they go. I like to take the time to look into every passing woman's eyes. Most ignore me. Some of them return the gaze. Very few smile. None of them stops.

Something catches my attention and I freeze, finding myself in front of the window of a cafe. Freshly drawn onto the window is a symbol I'd seen just a few days ago. A Yin and Yang with skulls instead of dots. *Be balanced or die.* Right on the other side of the window sits a young woman with a faraway smile on her face. Even though I'm standing right in her field of vision, she doesn't notice my existence. It's one of those tiny cafes with only four tables which is full of books and paintings squeezed together on the walls. The woman seems to enjoy her daydream, while drinking some sort of hot beverage despite the high temperatures of the Barcelonian summer.

I hesitate. There is something unusual about her, but she has the gaze of someone that just meditated for hours. For some obscure reason, I find that scary. I guess I must have that look from time to time as well. I try to meditate half an hour every day. For me, part of it is about making myself more self-aware. But there is a long way to go.

Let's do an experiment. Put your hand behind your back and make a peace sign with your fingers. How do you know that your fingers did what you told them? You just know, right? That's because you have body-awareness. You know what your body parts are doing without needing to see them. You need body-awareness all the time, to do even simple actions, like walking, picking something up or sitting down. Body-awareness lets healthy people do amazing things, like playing an instrument, dancing, acrobatics or yoga. However, there are people whose body-awareness is strongly impaired. They don't know what their body parts are doing when they can't see them. There are cases of people like that who wake up in the night and accidently touch themselves. Not knowing what their own body parts are doing, they mistake themselves for someone else and start defending against a non-existent attacker, beating themselves up.

That is what it is like with our minds. We have no awareness of what our minds are doing. So we beat ourselves up mentally. That is why I meditate, to become aware of what my mind is doing. To stop beating myself up. Life can be so much better with a little more awareness and less self-beating. It frees your time and energy for family, friends and activities which make you happy. And that is just the start. I believe that gaining full awareness of the mind is the next step in human evolution.

Following a spontaneous impulse, I enter the cafe and take a seat at the young woman's table, directly in front of her. I look into her eyes. A small part inside me hopes to make her instantly climax. She has bright eyes roofed by intense dark eyebrows and wears a pullover which is probably one of the thousand creative creations of her long-dead grandma. Her hair is colored in a kind of red, which seems to significantly contribute to the brightness of the cafe.

The young woman looks right back at me. She doesn't seem surprised, or annoyed. Slowly she reaches for some huge glasses on the table, while keeping eye contact as if her life depended on it. In a careful movement, she places them where they belong. Based on the thickness of the lenses I figure she has about the same level of vision as the average corporate banker. The glasses cover half her face, but they can't hide the expression of amusement that emerges as she takes me in.

"At first, I thought a giant ape with a questionable taste in hairstyling was sitting in front of me. Now I am disappointed to see you are just a damn hippie."

I suppress a slight feeling of being caught. I guess I do look like a hippie. But I do not see myself as the stereotypical hippie who lazes around all day, smokes weed and never washes himself. I value authenticity, connection, love and peace. I might be a hippie by heart and apparently also by looks. But that's it. Nothing more.

Why am I justifying myself? I let it go. I feel proud because it works. This young woman inspects my black hair and my dark, trimmed beard. I know that I look wild and groomed at the same time, even though that may sound contradictory. That's on purpose. But don't tell anyone.

"Save your disappointment for when the hippie doesn't ask you out." I act as if I am offended while letting her know I am cool with a little smirk.

"That's fine with me. I am more the ape type anyway." She covers a smile. "But that problem is solved quite easily." She takes off her glasses again. "Now ask me out ape!"
I can't help it. I laugh.

"You are completely nuts. What's your name?"

"Alice."

"Bob."

We shake hands as if this is a formal business meeting.

"What are you doing here? Dreaming about world peace?" I try to joke.

"World peace? Stick to your role ape! World peace is what hippies dream about. You dream about bananas. I, as a nerd per definition, am dreaming about quantum physics."

Alice says this as if it should be evident to anyone. *What an arrogant bitch*, I think. "That's really interesting," I say.

In fact, some part of me does find it interesting. It's the second time in a week that life confronts me with quantum physics. *Is this a sign?* Perhaps Alice can tell me something about this entanglement thing, and the weird cat everyone is so obsessed with. Maybe I'll find out how to repeat my experience with Samantha. A part of me likes the idea of being able to induce orgasms everywhere I look.

"Interesting yes, but once you enter the rabbit hole of quantum physics, there is no return. Never. It might become the best, or the worst day of your life."

Her voice sounds too deep for her appearance. Also, there is something cold about it. Despite the heat of the Barcelona summer, I shiver.

"My worst day because I have to listen to your voice?"

"Possibly. I love to talk about quantum physics. And you might feel as stupid as you look."

I like witty women. They electrify the sacred dance between feminine and masculine energies.

"That's okay. Feeling stupid means that I am learning. Feeling smart means stagnation, my little nerd." I smile.

"Excellent attitude, Bob. Can I buy you a beer?"

Not the worst idea on a hot summer day.

"Sure."

I am not so sure if I like her. For a nerd, she is quite cool, but she also has something that's new to me. I could call it energy, but perhaps a more accurate description would be controlled unpredictability. She seems to know exactly what she is saying and why she is saying it. But despite her sureness, I have a hard time anticipating what comes next.

Alice looks for the waiter who is playing with his phone in the opposite corner. In this tiny cafe, that means he is only two meters away and probably hears every single word we say.

"I think you two just won the prize for the weirdest couple to ever sit here," he says.

"And you are winning the award for the most indiscrete waiter on the planet! Maybe you could distract yourself by getting us two beers?" Alice replies. She then looks at me. "It's funny. Why do humans behave so differently when they feel observed?"

Why would she ask something like that?

"I guess it's because of self-concern. We see and judge ourselves through the eyes of others, and we are obsessed with the image we believe others might have about us," I respond.

Alice frowns. "That's stupid on so many levels."

"It is. But that's what we do. All of us."

I wonder what Alice thinks about my intentions. Does she think I want her? Thoughts identified. *Stay present Bob.*

Alice smiles while lifting her beer. Tantra is not just about sex. It's about letting go, accepting what is and allowing things to happen. There is no way to control this interaction anyway.

"Okay, let's get started! Are you sure you want to hear about quantum physics?"

"Sure," I say, thinking about my perspective on life. "Isn't it all just vibrations? Everything is energy, right?"

Alice looks at me as if I had just told her the Earth was flat. "It's true. Everything is energy. But if you can't say what kind of energy you are talking about, it has no meaning. It's as if you said nothing at all," she says sharply. "In science, there are many different kinds of energies, and each one is precisely defined."

I obviously triggered a sensitive spot in her. But I appreciate her directness. I prefer that to people who smile at me and nod while thinking to themselves that I'm an idiot.

"You think that only science can explain the world?" I say intending to provoke her.

"Yes. I think that all good explanations are scientific."

Damn arrogant scientists.

"That statement you just made, is that a consequence of science?"

Alice slowly leans back, while probably sensing that I am up to something.

"No, I suppose it's not."

"Then according to itself, it can't be true. The theory that all good explanations are scientific refutes itself because it itself is not derived from science."

Bam. I know I got her. Even failed philosophy classes sometimes pay off. Alice will deny that I am right, but I am tasting the sweet flavor of victory anyway. Alice slowly takes a sip of her beer, her brow ever so slightly furrowed. Then she looks at me surprised.

"You are right. It really does rule itself out. It follows that there might be good explanations beyond science. Thank you for updating my map of reality."

I did not expect that. I am impressed by how rationally Alice responds to a good argument. She doesn't take it personally, and she doesn't seem to have a problem admitting that she was wrong either. I like that.

"I have never thought about it that way," her voice sounds warmer, and I sense increased respect towards hippies. And probably apes.

Yeah! I genuinely dislike the idea that science knows everything.

"You know I have experienced some bizarre things in my life," I say. "Things I can't explain with science at all. And if I want to talk about them, I simply use the word energy. It's true, I'm not sure exactly what it means. But I don't care if you nerds don't want to share the word. I don't need to prove anything. For me, it's enough if I believe it. I don't have to justify myself to anyone."

"I understand. Still, that sounds like a justification to me." *Damn it*, she is right. That's precisely what I am doing. *But why?* Because there is doubt. A part of me wants to believe that yogis can fly. That telepathy exists. That humans can live from light only. That I can induce orgasms by looking. But maybe my mind is tricking me. My mind is an expert at that.

"Okay fine, maybe you're right. So quantum physics is more than vibrations then? You'll have to explain to me how it works."

"It would be my pleasure. Ever wondered what an atom fundamentally is? What light is? Or how atoms talk to each other?"

Atoms talk to each other?

"No. What are those atoms talking about?" I picture two atoms insulting each other. One of them is a nerd, the other one, a hippie.

"They have a serious conversation about what they know about each other. Since atoms don't like to keep secrets, we would probably call that gossip. Nothing is ever forgotten. Information cannot be created or destroyed, just passed on. And that's what they do. But their conversation style is extraordinary because they behave as if they live in many different realities at once, and they talk to themselves a lot. If people could listen to that shit, our psychiatric facilities would be full of those atoms."

"But they are, right?" I say. *She is definitely nuts.*

"Right, they are, because those damn atoms are everywhere. What they do and how they talk to each other completely underpins our perception of reality and even the existence of life itself. And what determines how atoms interact and chat? Quantum physics! Everything is built on top of it. Everything in this universe is made of tiny particles, and they all live by the rules of quantum physics. But you know what the strangest thing is for me? The atoms in our bodies are exactly the same as those in any tree, plant, animal, rock or star. And for some reason, those atoms organize themselves into a human, who is now trying to understand atoms."

I imagine a bunch of atoms coming together and then asking, "what the hell are we?". And that's crazy because that's what they actually do. But why? It reminds me of restless meditation retreat addicts, wanting to find their true nature by meditating until their legs hurt while singing mantras until their throats are sore. I was one of them. It started five years ago, after the accident.

From time to time I still talk to my son. David. Today would be his tenth birthday. Human lives are not linear. They are not predictable in any way. Sometimes small seemingly insignificant choices have enormous consequences. Often, it's just pure chaos. Starting a conversation with a stranger on an airplane might change your life forever. Not paying attention to the little things might destroy your family. You can't predict what terrible or beautiful consequences a single prolonged eye contact might have. But we still feel responsible.

From the day my son died, the claws of guilt gripped my heart and since then, never stopped squeezing. In the beginning, it was unbearable. It still is, sometimes. I fell into a sticky hole from which I thought I would never escape. But with hard work and time, and quite radical shifts in perspective, it got better. The philosophy of tantra helped me a lot and I told myself that I'd grown past it all. Now I am going to combine tantra and quantum physics to find my well-deserved peace. I want to free myself of the past. Completely. "Atoms are trying to understand atoms? That just took me on a trip," I say honestly, while trying to return to reality. I'm getting a sense of why some seekers are obsessed with quantum physics. If it's really the foundation of everything, it might conceal an important message. A message so fundamental that it undercuts our suffering at its deepest roots. "But, why did you say, '*if* people could listen to the conversation of atoms.' Can't we listen to them?"

Alice: "No, we can't. And this is tricky. It is as if the atoms are doing a theatre play, the theatre play of quantum physics. We are in the audience trying to watch them on the stage, but the whole time the curtain is down so we cannot see what is going on. Then if we try to look behind the curtain, even just the slightest peek, the quantum theatre stops.

The quantum actors only play if nobody is watching. When we look, the actors pretend to do something else."

Me: "Sounds like the quantum actors behave like teenagers watching porn or smoking weed. As soon as their mom knocks on the door, they stop and pretend to do their homework."

Alice laughs: "Looks like somebody has some experience there."

I think about smoking a cigarette. Although I stopped a month ago, my brain doesn't seem to know that yet.

Me: "Yeah, I must have read that in a book or something. So the quantum world is fundamentally obscured? Nature is playing with masked cards? Isn't that a bit frustrating for scientists?"

Alice: "It is. But we are not completely lost. We can still see the *consequences* of the obscured quantum theatre. What happens behind the quantum curtain dictates what is happening in our normal, classical reality. The classical stage. And by making careful observations, we can get an idea of what is going on behind the curtain. We can't see the teenagers smoking, but we can smell the weed."

This might become interesting. It sounds like those atoms could use some shadow work.

Me: "What did you mean by our 'normal, classical reality'?"

Alice: "The reality we seem to live in, the reality we perceive. It's the reality where things seem to exist independently from each other, where things seem to be certain. Where objects seem to have only one position and only one velocity, not many at the same time. We call it 'classical' because that's how scientists used to think the world worked before quantum physics came along and changed everything."

Me: "Ok, so instead what is going on behind that curtain?"

Alice slowly leans toward me and looks at me seriously.

Alice: "Behind that curtain, we find a strange world. A world where a possibility is as real as a certainty. A world where contradicting possibilities dance and influence each other. The intriguing thing is that we can model this dance extremely well with mathematics, but for almost a century we have been struggling to understand what it all means for our concept of reality. It goes so far that many scientists refuse to talk about it. This approach is so common that there is even a name for it: 'shut up and calculate.' In those rare cases where people do talk about it, they usually disagree."

Me: "That's incredible. So the mathematics is unambiguous, but there is room for interpretation?"

Alice: "Exactly. There is no controversy about the predictive power of quantum physics. Quantum physicists are every day testing its predictions, and so far, it hasn't been wrong even a single time. Even after thousands of precise experiments examining each tiny detail, the theory unshakably holds. It's remarkable how damn well it works.
And it's also incredibly useful. Without our understanding of quantum physics, we wouldn't have been able to invent today's computers, modern cameras, smartphones, lasers, energy-saving LEDs, solar cells, GPS, MRIs, the internet, and thousands of other almost magical things."

Well, I am not a fan of computers and smartphones. I don't even own a smartphone. It makes me sad seeing all those people in the metro getting their soul sucked into a thing in their hand, while it makes them play thumb wrestling against some invisible opponent. If the pace of evolution kept up with our use of technology, we would all have at least ten thumbs by now. My phone is a Nokia 3310, and its battery lasts for a week. I use it to call people. And to throw at people. It's survived a lot of shit. I think the world would be a better place without smartphones. I am a big fan of solar power and energy-saving LEDs, though. And I have a weakness for lasers. Maybe because last summer I had my eyes lasered. I should recommend that to Alice but based on her glasses that would probably require a laser the size of the Sagrada Familia. For me, it worked amazingly well — goodbye annoying contact lenses. So I have to give this theory some credit. But it sounds strange. Everything is obscured by a curtain we can't look behind? *A dance of possibilities?* What does that mean for my everyday life?

I wonder what Alice is thinking about. Based on her dreamy smile it seems she likes drifting off into her inner world as much as I do into mine. I like those vibes. It's rare to find someone that you can enjoy stillness with, without falling victim to the urge to say something meaningless just to avoid that awkward feeling silence can create in us.

Me: "So do *you* have an idea about what is *really* going on behind the quantum curtain?"

Alice: "Sure I do. If you want, I will take you for a ride. We will build quantum intuition and explore the mind-blowing consequences of a theory which seems to describe how nature fundamentally works."

I hesitate. It sounds fascinating I have to admit. But will this knowledge improve my life? Or am I wasting my time? Understanding how nature works? Is that even possible? I think about the cat. I think about entanglement. I think about Samantha, the quantum healer. I take a sip from my beer and decide I am in.

"Okay, Alice. I am in."

Alice: "Cool. Let's start with **wave-particle duality**. Ever heard of that?"

"No," I say, remembering Descartes mind-matter duality from my classes in philosophy. *Just listen, Bob. Just listen.*

Alice: "Okay, imagine a wave. For example, a water wave at the beach. It's spread out over a large area, right?"

Me: "Right."

Alice: "Now, imagine a tennis ball. Unlike a wave, a tennis ball is not spread out. You can tell exactly where it is. So, here's the question. In your everyday life, have you ever seen something being a wave and a tiny tennis ball at the same time?"

Me: "No, of course not."

Alice: "Exactly. To us it appears that something can't be spread out and not at the same time. In our classical map of reality, those two properties are exclusive. But many rigorous experiments have shown that light, electrons, atoms and even complex molecules[3], behave as waves if left alone and if we try and see where they are, they suddenly act like little tennis balls. They appear to have both properties. That clearly contradicts our everyday intuition. It is as if you went down to the beach and you could hear the crashing of the waves, but every time you looked at the sea you just saw a little ball of water at some random point on the seashore."

That confuses the hell out of me. I order some Catalan house wine to try and lever open my mind.

Me: "Okay, that's weird. So what are those atoms now? A little tennis ball or a wave?"

Alice: "Using the word 'or' shows that you are trying to force the world on to a map of reality in which those two properties can't coexist. Forget everything you know about how things should be, and this will be a fun and enlightening ride, my little hippie."

That's precisely what I wanted to hear. A new way of seeing the world. A new perspective to wash away my troubles.

---

[2] Nairz, Olaf, Markus Arndt, and Anton Zeilinger. "Quantum interference experiments with large molecules." *American Journal of Physics* 71.4 (2003): 319-325.
[3] Eibenberger, Sandra, et al. "Matter–wave interference of particles selected from a molecular library with masses exceeding 10000 amu." *Physical Chemistry Chemical Physics* 15.35 (2013): 14696-14700.

Me: "You sound like my stepdad the first time he gave me acid."

Alice: "Well, there are some parallels. Both take you out of the bubble that you believed to be the only reality."

Me: "Right, we do all live in a bubble. But what has quantum physics got to do with that?"

Alice: "Thinking about quantum physics has some serious side effects. When you study quantum physics, your bubble starts cracking. You start to wonder. You become humble about what you know about the world. You start questioning your map of reality and stop mistaking it for reality itself."

Me: "You keep saying 'map of reality'. I know the term from my meditation circles, but I've never heard a nerd use it before. What does physics have to do with that?"

Alice: "Look at this glass of wine. What do you perceive?"

Me: "A glass of wine."

Alice: "Yes. But is there really a glass of wine? Is it objective? For example, would an elephant see a glass of wine as well? Fundamentally, a glass of wine is just a bunch of atoms. Light bounces off those atoms and enters the small hole in the center of your eye. Inside, light is converted into electricity from which your brain generates an image of a glass of wine. Only a tiny fraction of the information available about the glass of wine makes its way to your consciousness. Your brain is creating a simplified map of reality by which you can live and make decisions. And this map is driven by millions of years of evolution, so as well as simplifying, it may also distort reality to help you survive life. This is not just about glasses of wine either, but any beliefs we have about the world, about other people and ourselves. They are all just maps of reality. And quantum physics shows us that some very fundamental beliefs we have about the world must, in the end, be wrong."

Me: "Interesting, I guess I'm more familiar with the maps we have about ourselves."

I catch Alice gnawing on the frame of her glasses while talking. I guess her map of reality of how things taste must also be quite distorted.

Alice: "I believe most conflicts on this planet exist because humans tend to mistake their map of reality for actual reality. By curing just that we could end dogmatism, irrational intolerance, relationship issues, and religious wars. Nobody could insist on their beliefs anymore. It's evident that the map of Barcelona is not the same as Barcelona. There is no real beach on the map. In the same way, we never perceive reality as it is, just our own personal map of it, which gets triggered by electrical signals coming from our senses. If people would understand that alone, they would stop fighting each other due to differences in their maps. Instead, they would want to learn from each other to improve and extend their personal maps of reality. That would be a huge step closer to world peace, motherfucker."

Alice smiles at me knowingly. She is thoughtful in a provocative way. Her making fun of me being a hippie pisses me off and excites me at the same time. *That's just your ego talking, Bob. Don't take it personally.*

Me: "World peace, motherfucker? You should give talks at hippie festivals."

I picture Alice as a quantum physics guru standing on a chair in front of her followers preaching quantum physics and world peace while insulting them as damn hippies. I have to laugh.

Alice: "Are you sure they can handle that?"

Me: "Yeah. For sure. Some hippies can be quite dogmatic, so a dose of quantum physics might be exactly what they need. So what is going on now with this wave-tennis ball thing?"

Alice reaches for her bag and pulls out a small piece of cardboard. As she holds it above the table, I notice two tiny slits carved near its center.

Alice: "Focus on the shadow cast by the cardboard. What do you see?"

Me: "Well. Just two small and blurry white stripes in the middle of a rectangular shadow."

Alice reaches into her pocket and takes out a small laser pointer, turns it on and shines it onto the slits. A complicated red pattern appears on the table. Multiple stripes. All next to each other.

Alice: "What about now?"

Me: "Now I see a bunch of stripes, like seven. But there are only two holes in the card. How is that possible?"

A laser focused on a double-slit creates an interference pattern on the table. Nerd note: slits and pattern not to scale, but it does work.

Alice: "That's because light behaves as a wave. Different parts of the light wave go through the two slits in the card. Then afterward these two parts spread out and mix together creating the complex pattern."

Me: "And why does that only happen with a laser and not with sunlight?"

Alice: "The light coming from the sun or a lightbulb is a bit different from the light coming from a laser. A laser is like a sea where the waves are very regular, all having the same distance between them and all heading in the same direction. Sunlight, on the other hand, is more like after someone just dive-bombed into a small pool, waves of all different shapes and sizes going in multiple directions. The regularity of the laser wave makes the pattern visible, for sunlight the pattern gets washed out."

Me: "Okay, so light is a wave. What's strange about that?"

Alice: "It's strange because light is not a wave. Light comes in little energy packets, which behave like tiny tennis balls. Just like your body is made up of tiny atoms, each beam of light is made up of miniscule packets of energy called photons. You can count them, and you can shoot them at things one by one. We now even have cameras sensitive enough to detect a single photon, and if we use one to see where a photon is, it always looks like they are in just one place, not spread out like waves."

Me: "Okay, that is strange. But are you nuts? Do you always carry a laser and cardboard around in case some random hippie pretends to be interested in quantum physics?"

Alice: "You have no idea, Bob. You are the fifteenth hippie today."

Me: "Not bad. You must be obsessed with this stuff. So you're saying that light acts like a wave and also as a particle. But surely it can't be a wave and a particle at the same time?"

Alice: "I'll give you a metaphor. Look at this unfortunately empty beer can. This is a three-dimensional object. Now imagine you can only perceive its shadow. If I hold the can vertically in the light, its shadow looks like a circle, right? But if I turn the can horizontally, its shadow looks like a rectangle. Now, what is the can? Is it a circle or a rectangle?"

A beer can's shadow can be both rectangular and circular.

Sounds like an extended version of Plato's cave.

Me: "Dude, it's neither. It's a can of beer."

Alice: "Yeah, exactly. A can of beer is three dimensional. The confusion only comes up when we try to work out what it is from its two-dimensional shadows. And the same thing happens when we try to make sense of atoms and light. Even our best measurement instruments can only perceive reality's lower dimensional shadows. Sometimes its shadow looks like a wave and sometimes like a tennis ball. From our limited perspective that seems like a contradiction, but this wave-particle duality is just revealing the limits of our imagination. It confuses a lot of people because we like to believe that we see things as they really are. But something being two exclusive things at once contradicts that belief, and we have to admit that we don't see things as they are. Wave-particle duality shows us that reality is much richer than our everyday intuition can imagine."

I am impressed by the clarity of her thought process. I think I'm beginning to understand what she is trying to tell me. While Alice talks, I watch her closely. She seems to enjoy our conversation, but there is something odd about her. I can't pinpoint it, yet.

Me: "That's a good metaphor. Reminds me of the elephant, where some blind people touch different parts of it, and they completely disagree with their description of what an elephant is."

Alice: "That's a very nice metaphor, too. But in the wave-particle duality *opinions* of different people are irrelevant. It's nature itself we disagree with. And guess who wins that argument?"

Me: "Nature. Nature always wins."

Now I feel on familiar ground.

Me: "So, to you, a map of reality is a complicated network of beliefs we have about the world. And you want to say the world might be very different from these beliefs? So in other words, reality is not how it seems?"

Alice: "It's even totally different. Our senses detect only atoms and light. If we are smelling, we detect different kinds of molecules. If we are hearing, we detect vibrations of atoms. If we are touching, we are detecting the presence of atoms. And if we are seeing, we are detecting light. All the information collected by our senses is converted into electricity. And from that pattern of electricity, our brains simulate our experience. We never perceive reality. Your whole experience is just a simulation according to a map you have in your brain. How can that even be close to reality?"

It's true. We do create our own worlds. Our brains are wonderful machines. They project vision, sounds, smells, sensations thoughts, the experience of having a body, the feeling of identifying with that body, and even whole relationships onto the screen of consciousness. And all that is generated just from atoms and light?

Atoms and light talk to our senses. Our senses convert the information into electricity. That pattern of electricity triggers the map of reality stored in our brain which then projects our experience on the screen of consciousness.

In tantric philosophy, there is Shiva, which is consciousness. It is the empty substrate, the canvas onto which the movie of life is projected. And there is Shakti, the cosmic energy, which is responsible for the movie. Atoms, light, thoughts, feelings, everything, visible or invisible are manifestations of that single cosmic energy. In the philosophy of tantra, there is only Shiva and Shakti. Nothing else. I wonder how arbitrary our simulated experience finally is. Isn't it possible to simulate a completely different experience from the exact same configurations of atoms and light out there? What is it like to be a bat? It's interacting with the same light and atoms, but its experience is entirely different. Isn't what our brain does almost wholly responsible for our experience? But isn't our brain itself just a configuration of atoms?

Me: "Okay, so atoms and light are behind everything, including human experience. But human experience is made out of thoughts, emotions, and meaning. That's a completely different world. So why do we even bother?"

Alice: "Since atoms and light constitute physical reality, I believe they can teach us a lot. Most importantly, they can show us where our intuition goes wrong. Wave-particle duality is just one of many examples."

Good point. I don't find it very surprising that our intuition fails when we try to understand nature at its deepest level. We just did not evolve that way because it was not useful for survival in the past. I imagine Alice trying to impress a saber-tooth tiger with her wave-particle duality. I wonder how she and her ancestors managed to survive until now. The caring umbrella of society, I guess. Without that, those nerds would be in deep trouble. But now things are different. Maybe we really need to deeply understand the structure of reality to advance or even survive as a human species. Maybe we need to expose our false beliefs. Maybe we need to radically improve our maps of reality.

I think about the metaphorical gorilla, chained to a small pole as a baby by someone cruel. While young and weak, it could not escape, solidifying a belief of being trapped. Later, after growing older and stronger, the gorilla could easily escape, but convinced that it is stuck, it never tries. At that point, the gorilla is exclusively trapped by its own belief. Perhaps this is the reason why many people find quantum physics so fascinating. After all, it's a deep dive into the mystery of existence. And might, at the same time, be a road to freedom.

The beliefs acquired in its childhood trap the old gorilla.

Me: "So what else do we learn from this wave-particle duality, other than that we are too stupid to get it?"

Alice: "Listen. At the start of the twentieth century people already knew how to shoot electrons at things. You could shoot them at a screen and see where they hit, just like tiny tennis balls. But then, just like we did with the cardboard and the laser, scientists shot electrons[4] at a screen with two slits cut into it. We call it the **double-slit experiment**. It's a very simple experiment, and it seems quite random, but if you understand its weirdness, then you will begin to grasp one of the deepest mysteries of quantum physics."

I remember that an electron is a tiny charged particle which is responsible for electricity. It's also one of the ingredients of an atom. No electrons, no chemistry. No chemistry, no biology. No biology, no humans. No humans, no Bob. What a sad world that would be.

Me: "Sounds good."

---

[4] Davisson, C. J., and L. H. Germer. "Reflection of electrons by a crystal of nickel." *Proceedings of the National Academy of Sciences of the United States of America* 14.4 (1928): 317.

Alice: "Okay, so what happens if you shoot electrons one after another at the slits? You can imagine the slits like two open doors beside each other in a wall. Behind the doors, there is a screen, and the electrons leave little marks where they hit the screen. So you fire electrons one-by-one, some make it through the doors and mark the screen, some don't. After firing thousands of electrons, you start to see a pattern. The question is, what kind of picture do the electrons draw on the screen? What kind of shadow does the wall with the two doors cast?"

Me: "Sounds pretty boring to me. I would expect that the electrons draw a fuzzy version of the doors on the screen. Some electrons hit the door frames and bounce in weird directions, so the shadow isn't sharp."

Alice: "Right, that's a good answer. But it's wrong. What we actually see isn't boring at all. At first, the electrons seem to hit the screen at random positions. But then something happens that leaves many scientists speechless even today. After shooting thousands of electrons at the screen, slowly a pattern appears. The electrons do not draw just the two doors. They draw a pattern of multiple stripes, all next to each other, just like I showed you with the laser and cardboard. And this can only be explained if we think of each single electron as being a whole wave."

Bullshit. I don't believe that.

Me: "A pattern? Even though the electrons went through the doors one by one? That can't be true."

Alice: "When I heard that the first time, I did not believe it either. But remember the beer can metaphor. Just like a photon, an electron is **not** a tennis ball and also **not** a wave. But it can behave like both. I had to study and think about this experiment for more than a year to really grasp its implications. By the way, the same experiment has been done shooting atoms or even super large molecules at the slits. And we always see the same pattern of multiple stripes. Of course, the electron experiment shocked the scientific community. And until today, nobody knows *exactly* what is going on there. The math of quantum theory predicts the behavior perfectly, but people still disagree on what is happening behind the quantum curtain and what it means for our understanding of reality."

I feel dizzy. This wave-particle weirdness extends to all matter? *And you can observe it?*

Me: "Ok, that's super strange. So what do you think is going on? What does it mean for our view of reality?"

Alice: "Give me a minute. First, let me add a tiny little twist that makes this experiment even more mind-blowing."

Me: "Sure."

Alice: "So scientists thought the following: a wave creates the striped pattern we see because it travels through both doors at the same time. The part of the wave that goes through the first door mixes with the part of the wave which goes through the second door."

I imagine two lonely doors in the middle of the ocean and a wave going through both of them at once. Yeah, waves can go through two doors at once. That's for sure possible.

Alice: "A tennis ball, on the other hand, can clearly travel through only one door at a time, making the striped pattern impossible. So what happens, if we, in some way, find out which door the electron goes through?"

Me: "Good question. I am impressed by those nerds. I guess most of them have a hunchback for a reason."

Alice: "Yes, as a nerd, it's actually absolutely obligatory to have a hunchback. Otherwise, nobody takes you seriously. Also, you must take all measures to avoid getting tanned. And most importantly, you've got to make sure you don't accidentally follow any fashion trend. Or you will instantly lose all of your credibility."

Me: "You must be the queen of reputation."

Alice laughs: "I am getting there. I've been working a lot on my hunchback lately. Thank you for noticing."

I love it when a woman laughs at herself.

Alice: "Okay, now back to the mysterious electrons. So what do you think will happen, if we find out which door each of the electrons goes through?"

Me: "Intuitively, I would say it should not make any difference to the result of the experiment if I know what is going on or if I don't."

Alice: "I agree, but wrong again. So they measured which door the electron goes through. They thought, now we're going to get you, little electron, now we will know what you are doing little squirt. And again, the double-slit experiment has been repeated many, many times with all sorts of objects: light, atoms, and even gigantic molecules. And the result is always the same. When you measure which door the electrons, or atoms, or whatever goes through, that damn striped pattern just disappears! When the path they take is measured, electrons, atoms, gigantic molecules and even light draw just two fuzzy doors onto the screen. Just like you originally thought. As soon as you observe what they do, they stop behaving as waves and seem to turn into tennis balls."

My head feels heavy. Even though this is fascinating, I think I need a break — time for a cigarette. *No, Bob, you stopped smoking.* Ok, a beer instead. Atoms are also waves, and light is also a tennis ball? Everything is both? Or better said, neither?

Me: "*What the hell!* That is what you meant before, right? By checking through which door the electron goes, we try to look behind the quantum curtain, and the electrons stop watching porn and pretend to do their homework. But does that mean that the electrons do fly through both doors at the same time when unobserved?"

Alice: "Well, that experiment suggests that something is flying through both doors at the same time. Or that something which goes through one door, knows about the other door. But in fact, we do not know for sure what an electron does if it's unobserved. We do not even know if an electron exists if unobserved. But we know that the quantum theatre behind the curtain is a play involving some sort of waves."

Me: "Okay, so tell me, how do you imagine the quantum theatre? What is going on behind the curtain?"

Alice: "Alright, listen carefully. Behind the quantum curtain an electron really is a wave. And that wave does exist. Like a wave, it goes through both doors at the same time and then mixes with itself."

Me: "Okay, but what kind of wave is that? I mean, the substance of a water wave is water. But what is the substance of an electron wave?"

Alice: "Its substance is possibilities. It's a wave of possibilities."

Possibilities? The world is made out of possibilities?

Me: "Possibilities of what?"

Alice: "Possibilities of being or doing something a certain way. For example, the possibilities of being at a specific location. Or the possibility of going through one door or the other. What is happening behind the quantum curtain is a play of multiple possibilities and not a play of a single certain reality. Behind the curtain contradicting possibilities happen at the same time. Those possibilities talk to each other as if they were all real. They mix and they interact, even though they are exclusive in the world we perceive. And this dance of possibilities then gives us these strange results, which we observe on the classical stage."

A wave of excitement is flooding through my chest into my fingertips, bouncing back, building up in my stomach, moving down my back and slowly subsiding in my feet. My body already knows that this is important news. It is possible that my son is still alive?

Me: "Does that mean that everything possible can happen?"

Alice: "It has to be allowed by the laws of physics. But those laws do allow a lot. For example, if you drink a beer, the possibility that a dinosaur appears out of nothing and eats your birthday cake does not play a significant role."

Me: "That's too bad. That would make my birthday so much better."

I don't like my birthday very much. It is weird. It feels so fake to me to get all this attention just because the earth circled the sun one more time.

Me: "So, how does this dance of possibilities look?"

Alice: "Like I said, it's behind the curtain, so we can't observe the dance directly. But we can still visualize it since the possibilities act like waves. And waves have a frequency, they vibrate. Unlike tennis balls, waves can cancel each other out. A tennis ball plus another tennis ball always equals two tennis balls. Light plus light can equal light, but it can also equal darkness. That's why two slits can create a shadow of multiple stripes with dark regions in between. In the same way, the possibility waves can add together or cancel each other out. This is how they dance with each other. If you want to see a dance of waves, just throw two stones next to each other into a still lake."

Vibrating possibilities which cancel and enhance each other? I wonder what that means for us humans? I see us as vibrating conscious beings. The vibration defines how we feel and how we feel defines the vibration. A high vibration means that we feel good. And if we feel good, we think positive thoughts and have creative ideas. We have access to more possibilities. Our circumstances are a reflection of our feelings. That's because feelings distort our maps of reality and our maps of reality create our experience. That's no surprise since the feelings and the experience are both generated by the same brain. And both are cosmic energy.

It's a feedback loop. If you feel good, you don't get distracted by those energy-draining spirals of negative thoughts and emotions. Your whole experience is affected. You act differently. You see all the opportunities life offers, and you have the energy to take the ones you want, making you feel even better. If you want something positive in life, the first step is to improve your mood. Then your life becomes magical, your vibrations start to resonate with the universe. And if the physical world is really made out of vibrating and contradicting possibilities, maybe we can achieve things beyond our wildest imagination. That's awesome news.

Me: "This is amazing. So nature can just cancel out the possibility that an atom is somewhere? But then how does nature turn those dancing possibilities into certainties? I mean, things really happen in the end, right? We only end up seeing one of the contradicting possibilities."

Alice smiles: "Excellent question. It seems to us that nature manifests only one of those possibilities on the stage of our classical reality. Looking behind the curtain is a way of causing this manifestation. That's the reason the wave pattern disappears once you know which door the electron goes through. The possibility of going through the other door is no longer there. So the possibilities can't talk to each other anymore, they can't dance. By looking, you seem to kill possibilities."

I can feel how Alice's passion is starting to infect me. She *does* love to talk about quantum physics. Even though she is weird as hell, I have to admit, her passion is attractive.

Me: "So behind our experience of life there is no objective reality? Just different possibilities talking to each other and then somehow manifesting on our classical stage?"

Alice: "Well. Behind the curtain is no *single* certain reality. There are many realities. And those realities are playing with each other as if they were all true."

*Maybe they **are** all true?*

Me: "And what determines which reality finally manifests? What exactly kills the other possible realities?"

Is there room for influence? Can I make life manifest a reality I like? A reality where my son is alive? A reality where I am the master of quantum tantra?

Alice looks pleasantly surprised.

"I'm impressed. You found the deepest mystery of quantum physics. It still causes nightmares for scientists. The only honest answer is that scientists do not agree on what happens to the other possibilities. Also, it seems that we can't control which possibilities manifest on our classical stage. It happens by chance. We can only predict probabilities. Possibilities which have a high probability manifest more frequently than those with a low probability. The most common interpretation assumes that those other possibilities, after doing their theatre play, just disappear. They call that **the collapse of the wavefunction**. This view goes back to one of the founders of quantum physics Niels Bohr and is called the Copenhagen interpretation[5]. But the Copenhagen interpretation can't explain how this collapse of possibilities happens and where those unrealized possibilities go."

On the one hand, the image that things manifest by chance feels liberating. But another part of me does not like it. I want to be able to choose which possibilities turn into my life. I want to be in control. I don't want to be a victim of chance.

---

[5] Heisenberg, Werner. *The physical principles of the quantum theory*. Courier Corporation, 1949.

Wait, who is saying that? Relax Bob. Let it go. That's your ego speaking again.

Alice: "But I don't agree with this view."

What?

Alice: "Quantum theory tells us exactly what happens to those apparently disappearing possibilities. But the implication is so mind-melting that the majority of physicists still deny it."

Me: "Alright, I'm interested. So what is happening to those possibilities?"

I feel a strange mix of hope and curiosity.

Alice: "Be patient. We will get there step by step. But it's worth it. It's going to explode your mind."

Damn it, Alice.

Me: "Okay. So how do we even know about these possibility waves? I mean how did we come up with this? How do we know this is not complete bullshit?"

Alice: "Good that you ask! This is where the Austrian physicist Erwin Schrödinger enters the picture. In the year 1926, he did something remarkable."

Schrödinger? Wasn't that the guy with the ominous cat everybody is talking about?

Alice: "This guy just came out of the blue and said: 'Tell me the energies of the quantum actors, and I can tell you the exact choreography of possibilities at any point of space and time.' He invented a simple equation which holds no matter what[6]. The Schrödinger equation. An amazing achievement. Suddenly we could look behind the curtain in extreme detail, using abstract mathematics."

I imagine myself in a theatre, the theatre curtain raising, revealing a group of dancers. All moving in weird mathematical structures. Everything guided by a single equation.

Alice: "And it goes on. In the same year Max Born came along and said: 'If you give me the choreography of the possibility waves going on behind the curtain, I can tell you the probability of any event happening on the classical stage at any point of space and time[7].' *Bam.* That was a revolution in science. Suddenly it was possible to calculate what happens behind the curtain and to predict what we will likely

---

[6] Schrödinger, Erwin. "An undulatory theory of the mechanics of atoms and molecules." *Physical review* 28.6 (1926): 1049.

[7] Born, Max. "Quantenmechanik der stoßvorgänge." *Zeitschrift für Physik* 38.11-12 (1926): 803-827.

observe on our classical stage. Both received the Nobel prize for their work, and a new golden era of research began."

Me: "Can this Schrödinger equation explain the double-slit experiment?"

Alice: "It describes it perfectly. You can use the Schrödinger equation to calculate the exact shape of the possibility-wave of an electron when it passes through two doors at the same time just by doing straight forward mathematics."

Me: "Does that mean that the electron arrives as a possibility wave at the screen and right before hitting the screen it looks exactly like the whole wave pattern?"

Alice: "Yes! Each single electron *is* the whole pattern. But at the moment the electron hits the screen, only one of all the possibilities manifests, and we just see a single little dot on the screen. It's impossible to know where that dot will be. Each electron appears seemingly randomly on the screen. But we can calculate the electron's possibility wave. And it tells us the probability of an electron making a dot at each point on the screen. Then by sending many electrons through the doors, we slowly build up the pattern of multiple stripes, which looks exactly like the electron's possibility wave."

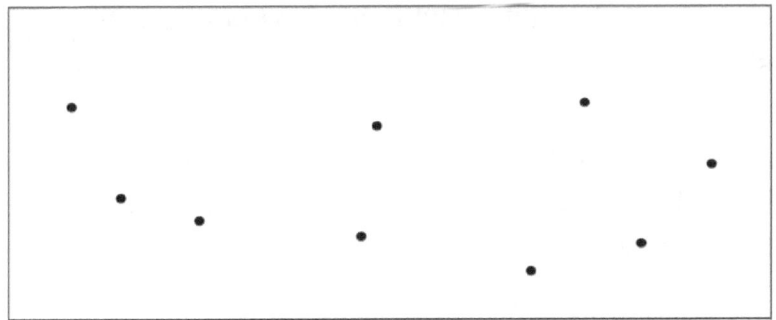

At first electrons appear to hit the screen randomly.

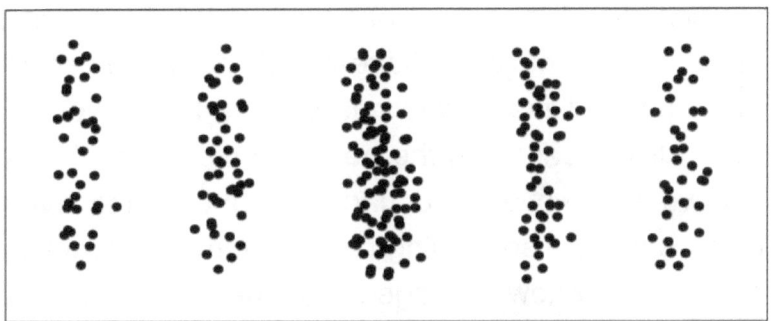

Many electrons create a pattern of multiple stripes.

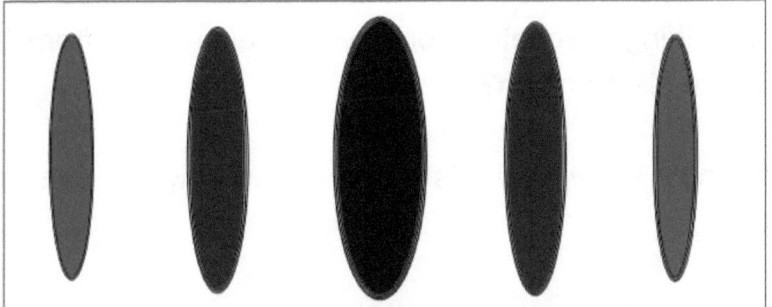

Schrödinger's equation predicts the pattern perfectly by describing each electron as a possibility wave.

Me: "That's beautiful. But I still don't get how this is all possible."

Alice: "Don't worry. Nobody gets it. We might get used to it, but nobody really understands it. From the human perspective, reality is weird."

Nobody gets it? What a strange world. I imagine going through two doors at once. How on earth would that feel? A physical world made out of possibilities? Possibilities that randomly turn into the solid real world as we perceive it. And those possibilities vibrate and behave like waves? This is what Alice meant with the beer can metaphor. Things can behave like waves and particles, but behind everything there is more – something beyond human perception. And still, we somehow managed to come up with an equation that seems to know precisely what's going on.

Me: "That makes me feel a bit better. Can you use the Schrödinger equation for anything else aside from telling where electrons go?"

If I only knew how this damn equation would change my life...

Alice: "Actually, this equation and a bunch of rules is basically all that quantum theory is. But there are almost no limits to it. It makes predictions about an incredible number of different phenomena. This equation predicts the strangest experiments in astonishing detail. But it also predicts a lot of weird stuff people still need to wrap their heads around. For example, the strange phenomena called quantum entanglement and of course Schrödinger's famous cat."

The only thing I hear is cat. I have never been so excited about a stupid cat in my whole life.

Alice: "The power of Schrödinger's equation also gave us a profound understanding of atoms. Previously, atoms were imagined to be like tiny solar systems, where the electrons circle a tiny nucleus, everything behaving like tiny tennis balls.

You've probably seen drawings of atoms just like that, they're everywhere. But it turns out that an atom like that cannot exist[8], those pictures are all wrong."

Me: "So without quantum physics there would be no atoms? We couldn't exist?"

---

[8] Olsen, James D., and Kirk T. McDonald. "Classical lifetime of a bohr atom." *Joseph Henry Laboratories, Princeton University* (2005).

I had no idea that even my life depends on quantum physics.

Alice: "Correct. But with Schrödinger's equation we can see that atoms do exist because they are made of possibility waves. For example, you know Hydrogen?"

Me: "Well, not personally."

Alice: "It's the simplest atom, just one electron orbiting a single proton. This is a possible shape of Hydrogen's electron calculated with Schrödinger's equation."

Alice draws a strange shape onto a napkin. Two blobs on either side of what looks like a donut.

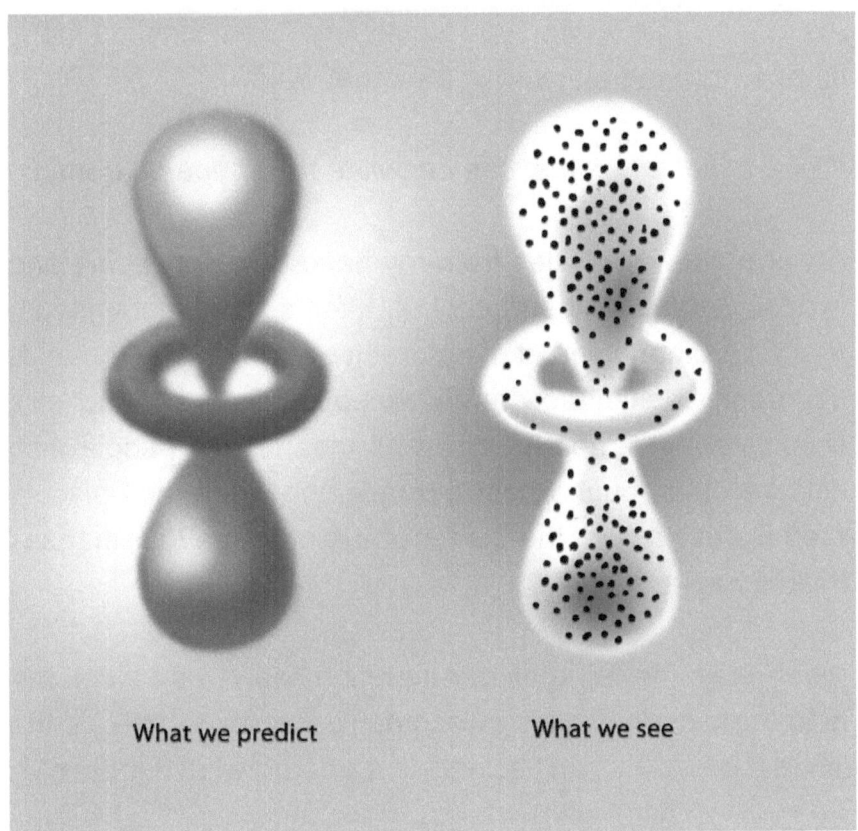

Left: The shape of the possibility wave of a single electron behind the curtain.
Right: If we look where the electron is, it manifests at a random point. Looking many times, we find the same shape as predicted by the Schrödinger equation.

Alice: "And this is what you would find if you checked many times where the electron in the atom was."

We, a bunch of atoms, can calculate the shape of atoms?

A warm shower trickles from my head down my spine and is absorbed by my heart chakra. It feels as if I just encountered a profound secret of nature. It's strange. Mathematics apparently describes nature exceptionally well. And still, scientists argue about what it all means. Did people also disagree that much about the meaning when they discovered Earth was not the center of the universe? Or that sex makes babies?

Me: "Okay, the Schrödinger equation allows us to look behind the curtain. That's awesome. So what, exactly, is the issue then? Why can't scientists agree on what it means?"

Alice: "Well, what happens to the other possibilities if only one manifests? Who decides *which* possibility becomes real? Are there hidden layers of reality we don't know about? These are the questions that people disagree on."

All told, I love the thought of the world made purely out of possibilities. When I imagine my future, that's precisely what I see. A landscape of possibilities where anything can happen. And maybe will.

Me: "Okay, so what is your approach to deal with this confusion?"

I must admit that I really want to hear what she has to say.

Alice: "It's very simple. Schrödinger's equation has been around for about a century, and during that time it has been tested and attacked over and over again. But every time it comes out unscathed. How can that be? Well, most likely because it's backed by reality. Thus, I take Schrödinger's equation dead seriously. And this includes all of its predictions, no matter how crazy they seem to us. Do not add anything or subtract anything. Take it as it is. Anything else would violate the principle of **Occam's razor**[9]. And don't worry, we will talk about those bizarre predictions in a bit."

Once again, I am impressed by the clarity of her thought. I remember the principle of Occam's razor from one of my philosophy classes. It basically states that the simpler a theory is, the more likely it is to be true. Simpler here means fewer assumptions. The reason is straightforward: each assumption has a probability of being wrong. If you need many assumptions to explain something, the probability of one or more assumptions being wrong dramatically increases with each additional assumption. So, by not adding anything to a working theory, you decrease the likelihood of being wrong enormously.

---

[9] Hugh G. Gauch Jr. *Scientific method in practice*. Cambridge University Press, 2003.

Me: "That makes total sense. But this is crazy! The universe is taking the piss. I mean, what was existence thinking? 'Haha! You know what? I am going to create a universe with conscious observers made entirely out of atoms. Then I will make them curious about themselves and the universe. But when they try to understand atoms, the very parts they are made of, I will mess with them by hiding everything behind a curtain. And the only way to see behind, is if they pass a math test by solving some equation.'"

Alice smiles: "That's true. But I quite like it that way. If the riddle of existence were easy, someone would have solved it already. Then there would be no more fun trying to understand, gain insights and learn more about the universe."

Interesting. I notice that Alice's cheeks are glowing with excitement. Alice seems to find happiness in the pure understanding of the physical world. I wonder why. Maybe I should introduce her to my physical world.

Me: "You seem to enjoy the process of understanding almost as much as knowing itself."

Alice: "I do. I love it. That's why research is my passion."

Me: "Awesome."

I feel happy for Alice. And envious at the same time. But that's okay. Out of the corner of my eye, I see the waiter approaching.

Waiter: "Guys! Don't you have something better to do instead of talking about atoms? I mean what's the use of that? Why do you even care? Go to the damn beach and get tanned like everybody else!"

Alice takes a long sip from her beer, lifts her head and looks at the waiter.

Alice: "You know, it's surprising that we are the only people in this cafe. With such a charming waiter."

Waiter: "Thanks for that honest compliment. Anyway, why do you need to understand everything? Doesn't trying to explain everything destroy the magic of *not knowing*? I prefer to live in a world with a bit of mystery."

I've heard this argument before, but I don't agree. I take a shot.

Me: "Well, look at it this way. Let's call the things we understand *knowledge*. For example, the way you use to make a cocktail is knowledge. Or the way to make a fire. Now imagine the sum of all knowledge as an island. The island of knowledge. The island is surrounded by everything we do not understand — the ocean of ignorance. Compared to the ocean, the island is tiny. Now, where does the magic happen? Where do we feel the mystery? Not in the center of the island, because everything is familiar there. And also, not far out in the ocean. Because if something is too far away out in the ocean, we cannot even perceive or imagine it. It's so alien to us that we cannot relate it to anything we know."

Waiter: "Yeah man, the mystery happens right at the beach! That's what I'm telling ya!"

Damn it. He's stealing my punch line.

Me: "Yeah. The mystery happens where we find unanswered questions, which is right at the border between what we understand and what we don't. Right there, where we kind of have a map, but something unexpected happens which we can't explain. Right there, at the beach, is where you find questions like 'Why do we love music?', 'What does quantum physics tell us about the nature of reality?', 'What does love without fear look like?' or 'How can we heal our destructive conditionings?' If we strive for understanding, explaining the things that seem mysterious to us today, the island gets bigger. And a bigger island means longer beaches. The border with the ocean of ignorance grows, and we experience new phenomena we can't explain. Then we can ask more questions. In that way, understanding and explaining things creates more magic and more mystery in our lives."

The island of knowledge surrounded by the ocean of ignorance. Mystery happens on the beach.

Alice nods. "And let me add that I would be devastated if we were to figure everything out someday. Not to mention, I would be unemployed. It's the process which is my passion — hanging out at the beach all day. Standing with one foot in the known and with the other foot in the unknown makes me feel the mystery of life every single day. I believe and I also hope that the ocean of ignorance is infinitely large."

> *"The most beautiful experience we can have is the mysterious. It is the fundamental emotion that stands at the cradle of true art and true science. Whoever does not know it and can no longer wonder, no longer marvel, is as good as dead, and his eyes are dimmed."*[10]
> **-Albert Einstein**

The waiter shakes his head. "So you're not going to the beach, because you're already at the beach? First, I thought you might be crazy. Now I know it. Any more drinks amigos?"

"A quantum wine please," I say, "it's red and white at the same time."

Alice laughs.

---

[10] Einstein, Albert. *The world as I see it*. Open Road Media, 2011.

"I see, now we are ready to talk about the damn cat," she says.

The cat!

Me: "Alright finally, the damn cat! Okay, what's its deal?"

Alice: "So first you should know that even though Erwin Schrödinger came up with the Schrödinger equation, he did not just completely swallow its predictions. As a good scientist, he was one of the biggest critics of his own theory and attacked its consequences mercilessly. He went as far as to invent a thought experiment to show the scientific community how ridiculous quantum physics was. Since this thought experiment involves a cat, it's now called Schrödinger's cat[11]."

That's confusing. In my experience, people try to be right. Why the heck would you attack your own ideas? But wait, what about Buddha? He attacked the idea he had about himself — the idea of being a person residing inside a body. According to him, this is an illusion produced by the brain, and he rejected it. I didn't know Buddha had what Alice would say were the makings of a 'good scientist'.

Me: "You're telling me that Schrödinger wanted to show that his own invention, the Schrödinger equation, didn't make

---

[11] Schrödinger, Erwin. "The present situation in quantum mechanics." *Quantum theory and measurement* 48 (1983).

sense? Why would he do that? Didn't he want people to believe in his equation?"

Alice: "Above all, Schrödinger wanted to know the truth."

> *"The scientist only imposes two things, namely truth, and sincerity, imposes them upon himself and upon other scientists."*[12]
> **— Erwin Schrödinger**

This completely contradicts my picture of scientists. In my view, they are arrogant people who think they know better than anyone else. And insult you if you don't agree with them.

Me: "I thought scientists were mostly dickheads. A bunch of people with big egos trying to show they are smarter than you."

Alice: "Unfortunately some scientists are like that. But their arrogance blinds them. And sooner or later they trick themselves into believing a theory that's wrong, immediately closing the door to the truth. It's exactly their arrogance that makes them fail. Good scientists are not only humble, but they also need to be aware of the tricks of their own mind."

---

[12] Schrödinger, Erwin. *What Is Life? the physical aspect of the living cell and mind.* Cambridge: Cambridge University Press, 1944.

Is this why I am here? Seeing through the tricks of my mind is something I've been trying to learn for years. Unfortunately, I have one of these brains that knows a shitload of mean tricks. Right at this moment, it's telling me that it would be the most fantastic idea anyone ever had in this whole universe to smoke just a single cigarette. Maybe the cat can help me out here. I would love to hear about it, but I feel an important question burning a hole into my throat chakra.

Me: "How can you tell if someone is committed to the truth or is tricked by their mind?"

I am thinking about my past experiences with people giving healing workshops. Or people selling stones, treated water or bracelets which are supposed to protect you from electro smog. I think about stories of fake martial arts practitioners. Yogis who claim they can fly or live from sunlight alone. Or people who claim they can access past lives and talk to your long-dead ancestors. All of them have some theory about why their stuff works. I've heard people explain that it has to do with information, resonance, frequency or morphogenetic fields. Is it possible that some of them are just fooled by their minds?

Alice: "That's super easy. Just thoughtfully disagree with their theory. The key is to be consistent and rational. If you are committed to the truth yourself, you will notice the difference very quickly. If they listen to you and want to know more about your view, they are interested in the truth.

They'll listen to you because by listening to you they can only win. If they find a flaw in your thinking, they gain even more confidence in their own view. If they don't find a flaw in your thinking, they've learned something new. Someone who is committed to the truth only gains by listening to someone who disagrees. However, if they get defensive, talk about conspiracies, tell you an anecdote or get offended, you know you are dealing with someone who is being tricked by their mind or blinded by their arrogance. The truth is secondary to them."

Wow. Over the years at various festivals and markets, purely out of curiosity, I did question the odd person. Very few listened to me. The vast majority responded defensively or told me some story about how their technique worked for a friend. One of them even ended up yelling at me, telling me that they didn't have to prove anything to me. I thought this was normal behavior for people who believe they know something. I thought they just wanted to defend the truth. But really, their minds just wanted to be right. I am not judging them. My mind is no better by any means. And even if it were, it's not their fault for having this or that kind of mind. I guess trying to confirm your beliefs is a normal human tendency. Alright, I'm starting to respect this Schrödinger guy.

Me: "Okay, that's pretty cool. So because he was committed to the truth, Schrödinger attacked his own equation."

Alice: "Yes, and he did it with a fascinating thought experiment. After almost a century people still disagree on what it means."

Me: "Tell me!"

Alice: "Alright, we start by putting a cat in a closed box. The box is well sealed so that no information can escape or enter. Inside the box is a device which has a 50% probability of killing the cat."

Woah, what is wrong with this guy? Was he raped by a cat in his childhood?

Me: "Like a guillotine?"

Alice: "The details of the mechanism are not so important. But in case you care, Schrödinger himself thought about a Geiger counter which detects the decay of a radioactive element. That detector is then attached to a hammer, which breaks a vial, releasing a poison as soon as it detects radioactivity. To keep it simple, let's just say that the decay of the element directly kills the cat. If the element decays the cat dies and if not, it survives."

Me: "Schrödinger must have had a real love of cats."

Alice: "Seems like it."

Me: "Okay, so now we have a really complicated machine to maybe kill a cat. Perhaps it would have been easier to just run it over? What exactly is the point? Why is everyone talking about this cat?"

Alice: "For a good reason. The point is that this cat-killing device relies on a quantum process, the decay of a radioactive element. You see, quantum physics allows two contradicting events to happen simultaneously. Just as a photon can fly through two different doors at once, a radioactive element can decay and not decay at the same time. So in that case, what happens to the cat? Since the element is both decayed and not decayed, then the cat must be both dead and alive. At the same time! All the stuff inside the box, the radioactive element, the hammer, whatever else and of course the cat turn into a single possibility wave. For a truly isolated box, Schrödinger's equation unambiguously predicts two coexisting realities. One in which the cat is dead and one in which it is alive. Those realities exist on top of each other. The cat is both dead and alive simultaneously.

And this is predicted by the same equation for which all other predictions turned out to be true."

Me: "A cat which is dead and alive at the same time? That sounds completely nuts. And how on earth can you insert a cat into an equation?"

Alice: "That's the power of abstract mathematics. You can invent symbols which represent anything you want and just put them into the Schrödinger equation and calculate what happens. For example, you can write $|\Psi_{cat}\rangle$ to represent the state of the cat. In the case the cat is alive and dead at the same time, it would look like this:

$$|\Psi_{cat}\rangle = |\text{alive}\rangle + |\text{dead}\rangle$$

$|\Psi_{cat}\rangle$ represents its possibility wave. It contains all the information about the cat. Scientists call that the **wavefunction**. The plus sign here means that both possibilities, a dead and an alive cat coexist. You can read the plus sign as 'and'."

Crazy. I had no idea that you can put a cat into an equation. That's not what I learned in my math classes. Well, at least I am pretty sure I did not. Ok, honestly, I have no idea.

Me: "Just to understand. Schrödinger's idea was to amplify the quantum weirdness of atoms to a living cat. So it became obvious how absurd quantum physics was."

Alice looks away. She seems to be concerned about something. Her vibes irritate me. *What's wrong with her?*

Alice: "That's right. That was exactly Schrödinger's point. He wanted to show how ridiculous the predictions of his equation were in order to show everyone that it couldn't be right."

Me: "Okay but well, a cat being dead and alive at the same time is completely absurd. I've never seen anything like that. Does that mean Schrödinger's equation is wrong for cats?"

Alice: "Not at all. History has shown that reality doesn't give a shit about what we humans find absurd. However, when it comes to attacking quantum physics, our dear friend Albert Einstein was very much on Schrödinger's side and, of course, loved his thought experiment. 'Your cat shows that we are in complete agreement,' he wrote to his friend Schrödinger. 'A wavefunction that contains the living as well as the dead cat just cannot be taken as a description of the real state of affairs.'[13]"

As Alice is ordering two more beers, I catch myself thinking about what our kids might look like. Giant blind apes? *You are living in the future, Bob. Relax.* But what's her plan anyway? Does she want to make me drunk? Something confuses me. She seems to like me, but her earlier intense eye contact has changed. She keeps looking away. *What is she hiding?*

Me: "Wow, even Einstein couldn't believe it."

Alice: "Yes, but that's no surprise. First of all, this all happened around 1935, and back then the Schrödinger equation hadn't been around very long. It hadn't been thoroughly tested. Nowadays things are different. We can create these so-called **quantum superpositions** in the lab. A quantum superposition is anything that is in a state of two exclusive possibilities at once. For example, we can put an atom in two different places at the same time[14], or make an electron spin both clockwise and counter-clockwise, or get a photon to fly in two different directions simultaneously."

Me: "Wow, that's just crazy as hell."

---

[13] https://erenow.net/biographies/einsteinhislifeanduniverse/21.php
[14] Kovachy, T., et al. "Quantum superposition at the half-metre scale." *Nature* 528.7583 (2015): 530.

Alice: "I know. But there is more. If the object in the quantum superposition is relatively large, we call it a **cat-state**."

Me: "Making fun of Schrödinger?"

Alice laughs: "Or honoring, as you wish. Physicists have made cat-states like an electrical current that flows in both directions along a wire at once[15][16]. Or a tuning fork consisting of trillions of atoms that both plays a note and does not at the same time[17]."

Me: "What? Are you telling me, that those things really exist? That people create them in the lab?"

Alice: "Yes! All those things do exist. They're all predicted by Schrödinger's equation, and now we can create them on demand."

Wow! It feels like the hair on my arm is trying to escape my skin. At the same time, heat rushes to my neck and forehead. *I just can't believe it.* I suspected reality was different from what I see, but that different? This exceeds my wildest dreams by lightyears. *Lightyears? You are turning into a nerd, Bob.*

---

[15] Friedman, Jonathan R., et al. "Quantum superposition of distinct macroscopic states." *nature* 406.6791 (2000): 43.
[16] Van Der Wal, Caspar H., et al. "Quantum superposition of macroscopic persistent-current states." *Science* 290.5492 (2000): 773-777.
[17] O'Connell, Aaron D., et al. "Quantum ground state and single-phonon control of a mechanical resonator." *Nature* 464.7289 (2010): 697.

Me: "So a cat, dead and alive, at the same time, could in principle exist? I mean, what the hell!"

Alice: "Yes. Even after desperately searching for almost a century, there is no evidence at all against a cat's ability to be both dead and alive at the same time. On the contrary, experiment after experiment confirms that it's possible to create cat-states with larger and larger objects. As a result, we now have much more confidence in Schrödinger's equation. There is no apparent reason why it would not also hold for the whole content of the box including the cat, the air molecules and anything else inside, other than that we humans would find it absurd."

Me: "But no one ever did this experiment with a real cat, right?"

Alice: "The problem is, that we are technically still very far away from being able to do this experiment with a real cat. You would have to build a box which completely isolates everything inside from the outside world. Not a single atom or photon would be allowed to escape because that would leak information about what's inside the box. Like looking behind the curtain, that would ruin the quantum theatre. And also, I guess it might be difficult to convince the ethics committee. You might find Greenpeace tying themselves to all the cats in the neighborhood."

Picturing a bunch of angry, clawing cats attached to the face of an activist makes me laugh. I do love all animals, and I highly respect Greenpeace. I feel slightly guilty for laughing. *Why do you feel guilty, Bob? Do you want to see yourself as a kinder person than you are? Accept and integrate your dark side Bob. Now you're judging yourself again. Stop. Whatever arises in consciousness has already been accepted. Otherwise, it wouldn't arise.* I take a slow deep breath, and I give the waiter a signal for two more beers.
I will show you, damn brain!

Me: "Ok, I see. We just imagine that we have such a box and think about its consequences. I like these 'thought experiments'. So what happens if we open the box?"

Alice: "As soon as we open it, we will find a cat which is dead *or* alive. Only one possibility manifests on our classical stage."

Me: "But before we open, the cat is alive and dead at the same time."

Alice: "If you take the Schrödinger equation seriously, yes. That's exactly what it predicts."

This is the weirdest thing I ever heard. *There is no way this can be true.*

Me: "I just can't believe that. I mean, isn't science just another belief system? It might still be completely wrong, no?"

Alice: "You are right, it is just a belief system. But it's a special kind of belief system as it's the product of a beautifully simple process. The process stays the same, but the belief system never stops evolving."

Me: "Okay, but it's still just a belief system, right? Why do so many people trust it?"

Alice: "The belief system produced by the process of science is like a house which survived earthquakes, thunderstorms, impacts of meteorites, volcano eruptions, napalm attacks, fire and of course the big bad wolf trying to blow it down. The scientific belief system not only survives all of these attacks, but it is also dynamically improved by them since merciless attacks are an essential part of the process. If you had the choice, would you rather live in a house whose resilience has already been tested over and over again, or would you prefer to move into the house right next to it, which was just built yesterday and looks a bit shaky?"

Well, if moving into that house means believing the beliefs it's made of, then yeah, for sure I prefer beliefs that have resisted a lot of attacks. I wish I could admit that. But I can't.

Me: "But can't science be disproven anytime?"

Alice: "What do you use to disprove science?"

Me: "Hmm, science?"

Alice: "Exactly. That is what I mean by saying science is a process. The thing is that by now many of our fundamental theories have been attacked so often and so heavily that we can feel pretty safe. Also, when we use those theories to make things, we see in our everyday life that they work extraordinarily well. Airplanes fly and almost never crash. The engine of a proper hippie van works for decades. We can instantaneously listen to almost all existing music. We can speak to people on the other side of this planet in real time, and we can insult a million nerds with the push of a button. Doesn't that make you wonder?"

Insulting a million nerds is very likely the most useful thing you can do with a smartphone.

Me: "And what about all those studies that are financed by some evil business, which only get published if the results fit their agenda? How can you ever trust science like that?"

Alice: "Okay, I see where you're coming from. In this discussion, it's essential to differentiate between science in our society and the scientific process. Do you know the difference?"

Me: "I'm not sure what you mean."

Alice: "Science in our society is far from optimal. It defines which results get published, who are the authorities, how peer review works. It allows publications that are politically, economically or ideologically motivated. There is money, prestige, pride, and status involved. And this, of course, is all highly questionable. Unfortunately, this creates confusion about what science really is, leaving a lot of space for pseudo-science quacks to occupy. And there are a lot, the internet is full of them."

Alice seems angry. I understand that, I am angry about that too. Although anger is not my favorite feeling, I enjoy our emotional connection.

Me: "Yeah! You probably know how easy it is to publish fake scientific studies. And they can be used to make anything people want to sell sound like it's backed by 'science'."

Alice: "I agree. However, the core process of science is still incredibly amazing. Even though it's surrounded by this broken scientific system, the process obviously works. We can't doubt that many important insights and almost magical applications have come out of it. Ever heard of electricity?"

Me: "Ever heard of atomic bombs, guns and child pornography?"

Alice looks at me indulgently.

Alice: "Child pornography, really? But once again, I get where you're coming from. Look at it this way. The scientific process is like a kitchen knife. You can use it to cook a delicious meal, but you can also slam it into your
or someone else's knee. It's not the kitchen knife's fault what we do with it. It's ours. The problem lies in human psychology and our own selfish intentions. Instead of blaming science we have to grow up."

Damn, she is right. Having lived in multiple communities, I know exactly what she is talking about. We might look like adults, but in each and every one of us lives a little child. When this child gets triggered by not having its needs met, even the most mature person can act like a complete idiot. The greatest lesson I learned in life so far is that we are all nuts. The only difference is that some people know it, and some don't. And those who don't are the most dangerous.

Me: "Okay. So explain to me how the scientific process works."

Alice: "Sure! The scientific process is a creative, evolutionary selection process, very similar to evolution. But instead of testing genes by trying to kill them, science tests ideas and explanations. And those ideas come from human intuition. The scientific process takes those ideas and attacks them with rationality and then tests them against nature. The beautiful ideas, which explain a lot of different things and survive a lot of attacks, turn into trustable theories."

Me: "Intuition? What has intuition got to do with science?"

Alice: "How do you think Schrödinger came up with his equation?"

Me: "I have no idea."

Alice: "He guessed it. An educated guess, admittedly. But still, in the end, it was just a really good guess."

He just guessed it? That's completely crazy. However, where do ideas come from? Where do questions come from? From where else than human intuition?

I imagine bubbles of ideas emerging from an ocean of intuition. But as soon as they start ascending, a slingshot-wielding nerd shoots them down one by one. Bad ideas pop and good ideas survive. Those who resist being shot by a whole army of nerds turn into knowledge.

The process of science. Ideas emerge from the sea of intuition while razor sharp rationality shoots most of them down. The surviving ideas turn into a trustable belief system.

That really makes sense. And this simple process brought us most of our knowledge and technology? *Unbelievable.* On the other hand, the process of evolution is simple as well and that even brought us hippies into glorious existence.

> *"All religions, arts and sciences are branches of the same tree. All these aspirations are directed toward ennobling man's life, lifting it from the sphere of mere physical existence and leading the individual towards freedom."*[18] **-Albert Einstein**

Me: "So science automatically happens if humans properly combine intuition and rationality?"

Alice: "That's right. The process made by combining the two creates a trustable belief system. So trustable that we used it to fly to the damn moon. And soon to Mars."

Fascinating. On top of giving us technology, is the process of science useful for anyone who is seeking? What are we seeking, if not the truth? The process of science will never tell us the truth directly. But it does act like a sculptor, taking away everything that's not true. And the remaining sculpture might be stunningly beautiful.

---

[18] Albert Einstein, Out of My Later Years: The Scientist, Philosopher, and Man Portrayed Through His Own Words

Me: "Okay I get it. Since Schrödinger's guessed equation survived all of those attacks, we have to take it and his cat seriously. So let's assume we have a cat which is dead and alive at the same time. That must be very confusing for the good old grim reaper, aka Death. I mean what is he supposed to do?"

Alice smiles at me before looking deep into her beer, as if it conceals the answers to all the questions haunting her dreams.

Alice: "I like your imagination. Let's bring Death into the thought experiment. So if the cat is dead, he is happy. Because obviously, Death is a fan of dead things. If the cat is alive, he is sad, because he has nothing to do. And since the cat is a possibility wave, Schrödinger's equation predicts that Death also turns into a possibility wave. Death is happy and sad at the same time."

Wow, what an equation! You can even insert Death. I picture Death lying on a therapist's couch, all confused because he is simultaneously happy and sad. His therapist is Shakespeare, and infected by Deaths confusion he thinks "To be *and* not to be?".

Alice: "Actually this is a very nice example of quantum entanglement. Death and the cat are entangled."

Entanglement? That's what Samantha, the quantum healer, was talking about after our tantra session.

I recall how she came into my tent. She was wearing a thin white toga with a black, Chinese looking pattern on the lower half. Below her collarbone, you could make out the top half of a flower of life tattoo, in its symmetrical perfection. It was obvious she was not wearing a bra.

She entered the tent with a posture so incredibly straight, it seemed as if her head was being pulled up by some invisible rope. If it weren't for the movement of her legs as she moved toward me, I would have been sure that she was floating. She sat down in front of me, without saying a single word. She did not smile, she just stared into my eyes. Pleasantly surprised by this subtle intrusion, I stopped what I was doing, took a deep breath and looked back.

Shakespeare said, "the eyes are the window to your soul," and in contrast to her almost angel-like appearance, her soul was as black as the burnt Palo Santo quietly filling the tent with its sacred odor. What a lucky day, after all, there is no turn on quite like a black soul. As our breathing synchronized, I could feel the sexual tension flowing from her eyes into mine, down my spine and knocking on the door of my sacral chakra. A door that is, of course, quite useless. Mine is always open.

But then she started screaming. *What are you doing, Bob? Stop daydreaming. You are talking to Alice. Here and now.*

Me: "Entanglement? What's that?"

Alice: "Entanglement is nothing more than connected possibilities. Entanglement is also a consequence of Schrödinger's equation. The cat is at the same time dead and alive and Death is at the same time happy and sad. But these possibilities are not independent of each other. They are connected. If the cat is dead, Death is always happy. If the cat is alive, Death is always sad. This is how you would write their entangled state:

$$|\Psi_{🐈☠}\rangle = |🐈\;☠\rangle + |⚰\;😈\rangle$$

Death and the cat turn into a single possibility wave. You can't divide this wave into a part that just describes Death and a part that just describes the cat. They become an unbroken whole. Again, the plus sign means that two possibilities coexist. The first possibility is the cat being alive and

Death sad. The second possibility: cat dead and Death happy."

An unbroken whole? It's not the first time I've heard that. Isn't Buddhism saying the same thing?

Me: "Okay. That's interesting. So talking about the cat and Monsieur Death as if they were individuals does not make sense anymore?"

Alice: "That's right. You can only describe an entangled state as a whole, you can't break it up into parts. But this phenomenon of entanglement had one of the smartest men that ever lived as an enemy. He hated this prediction of Schrödinger's equation and argued against it for decades. He even called it 'spooky'. This man was none other than our dear friend Albert Einstein."

Me: "He really didn't like Schrödinger's equation."

Alice: "Yes. You have no idea how viciously this poor equation has been under attack. And still is. But that's a good thing. As I said, those attacks are an essential part of the process of science. And Einstein deserves a lot of credit because the way he attacked Schrödinger's equation gave us a much deeper understanding of quantum physics. He proposed a new thought experiment. He said, well, if entanglement can really exist between different objects, then I could just move one of the objects, like, for example, Death to the

Andromeda Galaxy while leaving the cat in the box here on Earth. Light from Andromeda takes about 2.5 million years to reach our planet, and Einstein's own theory of relativity says no information can travel faster than the speed of light. So Death is in the Andromeda Galaxy, happy and sad at the same time, while here on earth, the cat is in the box, dead and alive at the same time. But, as we said before, Death and the cat are entangled, their possibilities are connected. And Schrödinger's equation doesn't care about the distance between them. The equation for the combined whole is still the same:

$$|\psi_{🐱💀}\rangle = |🐱\ 💀\rangle + |🐈‍⬛\ 😀\rangle$$

Until there everything is fine. But now, here on Earth, suppose you dare to open the box. Schrödinger's equation says there's a 50% chance the cat jumps out of the box alive and a 50% chance it's dead. Let's say you find a dead cat. What happens now? Immediately, the combined state of Death and the cat appears on the classical stage and changes to

$$|\psi_{🐱💀}\rangle = |🐈‍⬛\ 😀\rangle$$

Instantaneously after the box on earth is opened, Death, far away in Andromeda, is now just in a state of being happy."

I imagine Death jumping up from the therapist's couch and declaring: "I am cured! I am suddenly just happy!" While Shakespeare realizes, "Ah, *not* to be, that's the answer."

Alice: "Even though Death is 2.5 million light-years away and no information can travel faster than light, it seems like Death's state of mood has been instantly affected by the box being opened on Earth. Einstein felt that was 'spooky' and he concluded that Schrödinger's equation couldn't possibly describe how things really work. It couldn't be right. He published a thought experiment along those lines in May 1935 in a famous paper[19]."

I must admit, I'm impressed by how these quantum physics people try to find the truth by making up thought experiments. That is pretty creative.

Me: "Well, that sounds like a good argument."

Alice: "It's a *brilliant* argument. Unfortunately, at that time, the argument couldn't be settled. Schrödinger's equation was extremely successful in predicting experiments, and that was all that counted to most scientists."

---

[19] Einstein, Albert, Boris Podolsky, and Nathan Rosen. "Can quantum-mechanical description of physical reality be considered complete?." *Physical review* 47.10 (1935): 777.

Me: "But why didn't they just check if entanglement was real? Why didn't they ask nature by doing an experiment?"

Alice: "Nobody knew how. Imagine someone gives you two boxes and in each of the boxes there's a coin. You want to check whether they are showing heads or tails, so you open the first box, and you find that it's heads. Then you open the second box, and you find it's tails. Then this guy gives you another pair of boxes, and another, over and again. You keep checking inside, the first coin seems random, heads or tails, but the second one is always the opposite of the first. What can you say about the coins?"

Me: "Sounds like they are entangled, like Death and the cat. I mean, how does the second coin know about the first coin if it's random?"

Alice: "Not necessarily. Einstein would say: Well, the sides of the coins were already determined before you looked into the boxes. The guy arranged them, so the first was the opposite of the second. There weren't any coexisting possibilities, the coins weren't both heads and tales at the same time. He would say the coins behave more like a pair of shoes. If the guy instead put one shoe randomly in one box and the other in the other box, you would also never find two right shoes or two left shoes. It's only because you didn't see him place the shoes that you can't tell which shoe is in which box, not because randomness is a fundamental property of nature. Einstein's view was that something was predetermining quantum states, something not included in Schrödinger's equation, that quantum physics was incomplete."

Me: "Okay, right. Actually, that sounds like a much more reasonable explanation than this spooky entanglement."

Alice: "Right. The million-dollar question was then: how do you experimentally tell the difference between entanglement and a pair of shoes? At the time it seemed like it was impossible to tell."

Me: "Okay, I see, so since the argument between Einstein and the Schrödinger equation could not be settled, scientists just told each other to 'shut up and calculate' and stopped talking about it?"

Alice: "Yes. And that was the situation for almost three decades. But then a bunch of hippies in California entered the game[20]. The hippies wanted quantum entanglement to be true. They thought quantum entanglement could justify their worldview where things like telepathy, psychic mind reading, soul healing, connections between souls, and so on were possible. So they sat in hot tubs, took psychedelic drugs and talked about quantum physics. Finally, almost 30 years after Einstein published his objection to quantum

---

[20] Kaiser, David. *How the hippies saved physics: science, counterculture, and the quantum revival*. WW Norton & Company, 2011.

physics, a brilliant physicist called John Stewart Bell came up with an ingenious idea[21]."

Damn. I would have loved to be one of those hippies. At sitting in hot tubs and taking psychedelic drugs, I am a real expert.

Me: "He found a way to distinguish entangled coins from shoes? Sounds like a really clever guy."

Alice laughs. It sounds freer than I would expect from a nerd. I like it. Has she practiced laugh-yoga?

Alice: "In a metaphoric way, this is precisely what he did. He found a way to distinguish connected coexisting possibilities from unconnected ignorance. The following story might give you a feeling for how he did that.
Lennon and Lovejoy are two rebellious students. They often come late to class. One day the class is taking an exam, and of course, Lennon and Lovejoy arrive too late to take part. As an excuse, they say they had a car accident with another car, and it was not their fault.
'Ok, you can repeat the exam if each of you can answer two questions,' says the teacher, 'but from now on you can't communicate anymore, and each of you has to write the answer on a separate piece of paper.'
With no other choice, they agree.

---

[21] Bell, John S. "On the Einstein Podolsky Rosen paradox." *Physics Physique Fizika* 1.3 (1964): 195.

'Okay,' says the teacher, 'the questions are: where exactly did the car accident happen and what was the color of the other car?'"

Me: "Pretty cool teacher."

Alice: "Indeed. If you want to know if someone shares the same pool of information (as connected possibilities do) you need to ask the right surprise questions. The answers to the questions are irrelevant. It is only important if the answers coincide or not."

Me: "Okay, so this John Bell found a way of asking surprise questions to reveal if possibilities are really connected."

Alice: "That's right. Exactly as the teacher finds out through these surprise questions if Lennon and Lovejoy really had a car accident."

Me: "Pretty smart. And who was right? Einstein or the Schrödinger equation?"

Alice: "In the last decades there have been multiple extremely delicate experiments that ask exactly the questions Bell came up with. In fact, in 2016, people from all over the world came together to do a giant experiment called The Big

Bell Test[22] in his honor. It, along with all the others, confirmed that connected coexisting possibilities exist. Entanglement is real and Einstein was wrong."

Me: "The hippies were right! So does soul connection and telepathy work? What about quantum healing?"

Suddenly, I am back in my tent. I hear Samantha screaming. Her eyes turn white. I don't know what to do. Is she in pain? One core idea of tantra is to let everything happen no matter what. Let go of control. For a moment that's pretty easy because I'm stunned anyway. So I do the only thing I am good at. I focus on how this situation makes me feel. But then she collapses to the floor, and I panic.

Alice slowly shakes her head.

Alice: "Sorry to tell you so bluntly, but unfortunately quantum physics gets misused a lot. A lot of mumbo-jumbo just sounds more believable if you put the word quantum in it. I am not saying that telepathy or spontaneous healing can't exist. But you can't explain those things with quantum physics, and the people who do are either charlatans who want to sell you their shit, attention seekers or tricked by their own minds. There is even a word for that: quantum woo[23]. It's defined as the justification of irrational beliefs by an obfuscatory reference to quantum physics. And there is a lot of

---

[22] https://thebigbelltest.org/
[23] https://rationalwiki.org/wiki/Quantum_woo

that. But you'll be pleased to know that there is a prediction of Schrödinger's equation ten times more mind-blowing than quantum healing and quantum telepathy combined plus aliens. Let's get into that after another wine. So far any questions?"

I do have questions. Samantha is lying on the floor. I quickly kneel beside her to improvise some terrible, uncoordinated first-aid, but her eyes are already opening. She moans. She smiles. She whispers, "Thank you." I'm still in shock. My head feels like I've just lost a fight against Muhammad Ali. What was that?

But then this smiling angel on the floor, who I've never seen before in my life, says three words which cut through my inner peace like no words have ever before. I doubt I'll ever forget them: "David loves you."

How on earth does she know about my son?

Alice: "What's wrong with you? Dreaming about bananas?"

I guess I drifted off into my thoughts again. What did Alice say? *Yeah, I do have questions.*

Me: "I dream about world peace, motherfucker. But yeah, what about human consciousness? If things manifest on the classical stage only when we look, is it because of consciousness that we don't see quantum effects in everyday life?"

Or maybe we do? Suddenly, I remember where I'd seen the Yin and Yang with skulls instead of dots before. The symbol which made me stop here in the first place. Samantha had it tattooed on the back of her left hand. That is some coincidence.

Alice: "Some serious scientists actually explored that idea[24][25]. But it turns out that consciousness is not needed to explain why the world appears classical to us.
Consciousness doesn't even appear in quantum physics. It doesn't appear in any physics."

Me: "Why not?"

Alice: "Because we don't have the slightest idea how a complex arrangement of atoms could possibly produce consciousness. That said, it might be that someday in the future we will find that consciousness is another prediction of Schrödinger's equation. But so far we have no clue."

Me: "What is your definition of consciousness?"

---

[24] London, Fritz, and Edmond Bauer. "La théorie de l'observation en mécanique quantique." (1939).
[25] Wigner, Eugene P. "Remarks on the mind body question. Symmetries and Reflections." (1967).

Identifying as a spiritual person, I've thought about these things a lot.

Alice: "It's almost impossible to describe it with words. If I were forced to, I would say it's the substance which makes experience possible. Like a TV screen, which makes a movie possible."

Me: "That makes sense to me. And it's so difficult since consciousness can't grasp itself as a concept. The Buddhists have a nice way of expressing that: A finger can point to anything, but not to itself. A knife can cut anything, but not itself."

Alice: "Interesting."

Me: "Okay, so you're saying that the fact we never directly observe contradicting possibilities can be understood without consciousness, right?"

Alice: "Yes. The reason is gossip. Gossip makes the world appear classical to us. And it's also the reason why quantum healing and all this other quantum woo stuff can't work in the way people claim it works."

What the hell has gossip got to do with anything?

Me: "What do you mean by gossip?"

Alice: "Gossip is just the spread of information. It turns out that for the coexisting possibilities to dance with each other, the information about the possibilities needs to be kept secret. If anything else not involved in the dance finds out about the possibility wave, it gets messed up, and the coexisting possibilities stop dancing. At the moment something else gossips about the possibilities, they become really shy. And each possibility just ends up alone in a corner of the dance floor."

Me: "Crazy. Those possibilities behave just like most humans. And that's the reason why the quantum theatre only plays its possibility dance if nobody looks?"

Alice: "That's right! By looking you extract information and that messes with the waves. Looking is a way of interacting with things. If you look at this glass of wine, a bunch of photons talked to the wine and then tell your eyes their story. But we humans are not special in terms of being able to 'look' at things. In the world, we live in, everything talks to everything. The glass of wine talks to light all the time and to a lot of air molecules which constantly bombard it. The light knows about the wine, the air molecules know about the wine."

Me: "But in some sense, we are special, no? Because me knowing about the wine can make it disappear."

I finish my wine in a single gulp. *Bam.* But what am I doing? I need to drink less.

Alice laughs. Despite all our drinking, she doesn't seem to be drunk at all. "Yes, in terms of what you can do to wine, you are definitely special. But quantum theory treats our bodies just as a bunch of atoms. For gossip to occur, it makes no difference if these atoms belong to a human or to air. It's just because of the non-stop ongoing conversation between everything. That's what makes the world we know manifest on the classical stage out of the underlying world of possibilities. In case you are interested, this mechanism is called **decoherence**[26]."

Me: "Amazing. So the 'and' joining possibilities turns into an 'or' because of gossip?"

That's not easy to imagine.

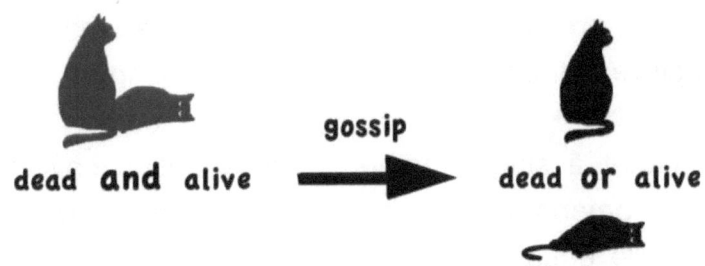

---

[26] Zeh, H. Dieter. "On the interpretation of measurement in quantum theory." *Foundations of Physics* 1.1 (1970): 69-76.

Alice: "That's right. That's the consequence of gossip. Turning 'and' into 'or' by messing with the waves."

Me: "How did they come up with that? I mean, how do we know this gossip stuff is true?"

Alice: "It's just another consequence of the Schrödinger equation. It comes out naturally if you take the spread of information into account. You can also measure the time it takes to change an 'and' into an 'or' in the lab. Again, Schrödinger's equation predicts these so-called **decoherence times** with creepy precision."

Me: "How can a single equation predict so many different phenomena?"

Alice: "Again, it's the power of abstract mathematics. The Schrödinger equation is something like a super-equation. Its solutions are not simple numbers, but instead an infinite number of additional equations. The deeper you go, the broader your predictions become. And the deep mathematical structure of the universe is surprisingly simple."

The fact that a single mathematical equation covers so many phenomena is really astonishing to me. How can it be? A single equation and so many different predictions. This is just beyond my imagination. Maybe it's worth getting into mathematics? But that's like learning an entirely new language. I guess I am a lazy hippie after all.

Me: "Okay, so all this gossip explains why we don't see co-existing possibilities and entanglement in everyday life. If you open Schrödinger's cat box/death trap, the atoms outside all gossip with the ones inside about whether the cat is alive or dead and pretty soon everyone knows. But I still don't get who decides whether the cat is alive or dead in the first place."

Alice: "Right, gossip explains very successfully why the world appears classical to us. However, it does not explain which specific possibility manifests. That still appears totally unpredictable."

Me: "Couldn't that still be caused by consciousness or some other weird mechanism we don't know about?"

Alice: "Sure. You can't entirely exclude that. Hey! Where the hell did you get that sandwich from?"

Me: "It just appeared out of nowhere. I guess that possibility just randomly manifested!"

Alice seems delighted. She does look hungry.

Alice: "Do you mind sharing?"

Me: "Of course, sharing is caring. Maybe in return, you could write down the Schrödinger equation for me? Even though I won't understand it, I just want to see it once. It must be quite complicated."

Alice: "It's actually pretty simple. The Schrödinger equation describing the state of the universe is:

$$i\hbar \underbrace{\frac{d}{dt}|\Psi\rangle}_{\text{change of possibility wave in time}} = \underbrace{\hat{H}\overbrace{|\Psi\rangle}^{\text{possibility wave}}}_{\text{sum of all energies}}$$

$|\Psi\rangle$ is the possibility wave of everything that exists. It contains all information about the universe. Just the region of the universe we can observe already includes hundreds of billions of galaxies, and each galaxy contains hundreds of billions of stars. And most stars have planets circling them, some of which are bound to be hosting life. $|\Psi\rangle$ contains your city, the chair you are sitting on, your body including all of its bacteria and of course your brain, which at this moment is struggling to grasp the overwhelming magnitude of $|\Psi\rangle$."

Wow. This is crazy. $|\Psi\rangle$ *describes everything there is?*

Alice: "Note that $|\Psi\rangle$ is one inseparable possibility wave. It's an unbroken whole. If you try to break it up into many smaller possibility waves, you will miss something."

Everything is just one single possibility wave? *"We are all one,"* comes to my mind. Is $|\Psi\rangle$ what enlightened gurus claim to have experienced?

Me: "What does the rest of the equation mean?"

Alice: "$\frac{d}{dt}|\Psi\rangle$ describes the dance of this possibility wave. In other words, how this possibility wave changes in time. $\hat{H}$ contains the sum of the energies of everything there is including their interactions. Schrödinger's equation says that the dance of possibilities is choreographed by the energies of the universe."

This is absolutely amazing. How on earth did Schrödinger guess this equation?

Me: "And what is this $\hbar$?"

Alice: "$\hbar$ is Planck's constant. It's a fundamental number which connects vibrational states with energy. For example, the energy of light is just $\hbar$ times its frequency."

Energy and vibrations are connected? I have no idea what that means, but the hippie in me likes the way it sounds.

Me: "And the small i in front?"

Alice: "**i** is something we call an imaginary number. It's the square root of -1. It's crucial because it makes $|\Psi\rangle$ unitary. Unitarity makes sure that the sum of the probabilities of all possible states in existence always equals 100%. Thus, there is always something."

It's astonishing how much information there is hidden in just a few mathematical symbols. To express the structure and beauty of the universe in a way that is so powerful and so elegant is incredible. I still can't grasp everything Alice is telling me, but for the first time in my life, I can sense the beauty of physics, the beauty of mathematics and the beauty of understanding in such a profound way that I'm left dumbfounded.

Me: "Wow. Just wow. I expected Schrödinger's equation to be much more complicated. And this simple equation predicts all the weird things we've been talking about?"

Alice: "Yes, it does. Actually, what I've told you so far is just a tiny fraction of its predictions. It also explains how magnets work. Why metal conducts electricity, and this chair doesn't. Why the sun is shining. Why a light bulb is bright. Almost all of our modern technology is based on that equation, and that's just the beginning."

Me: "And you said that not a single prediction turned out to be wrong? That's really hard to believe."

Alice: "That's right, not even one. It makes you suspicious, doesn't it? Suspicious that quantum physics is really how nature works. That it might really be a deep, fundamental truth."

I imagine the equations of physics governing the behavior of all matter, including us. Is life just unfolding and being guided by equations? Do these equations know everything? But where do they come from? God? And what about consciousness?

Me: "But what about the classical stage? You still haven't told me how this equation describes which possibilities manifest and which don't?"

Alice slowly leans over the table and looks at me seriously. A bit too seriously for my taste. I sense that she is struggling with something.

Alice: "Now it gets crazy. Because it doesn't describe manifestations at all. The classical stage idea was invented by a group of scientists in the early days of quantum physics. It's part of the Copenhagen interpretation of quantum physics. These guys knew that they only ever saw one possibility manifest, but they weren't sure how to deal with the remaining possibilities. So, they just said: There *must* be some mechanism which randomly favors one of the possibilities and kills the other ones. And that magically picked possibility then manifests on our classical stage."

Me: "So the classical stage was added to the theory? But do scientists believe in it?"

Alice: "Most of them never question its existence because it feels right. For this reason, it's been taught in most universities for almost a century. Even worse, many don't care, just going on with the 'shut up and calculate' mentality. Basically, the classical stage was invented because we tend to want the way we perceive reality to be the way reality really is. Just like before, when people were sure Earth was the center of the universe, we want our version of reality to be special. But this is not how reality works, quantum physics doesn't care how we perceive it."

I love it when my perception of reality is questioned. I am constantly searching for a perspective on life which maximizes my happiness. Or reduces my pain. I've had many shifts over the past few years. But often I realize later I was tricked. Tricked by my own mind. But this time it will be different. This time I will be moving into a completely crazy but resilient house. A house that survived bombardments from all directions. This time I will feel safe.

Me: "So you're saying that there is no classical stage? There is no curtain?"

Alice: "It is not me that's saying that. The Schrödinger equation is saying that[27]."

Me: "Then why did you tell me all this curtain stuff?"

Alice: "Because from our limited perspective it looks as if there is a curtain. As if there are manifestations. And in some sense, there is a curtain, but this curtain is completely subjective and therefore an illusion."

---

[27] Everett III, Hugh. "The theory of the universal wave function." (1956).

Me: "I don't understand what you mean."

Alice: "We will get there. Let's do another thought experiment. Let's get back to the cat. Imagine that you are standing in front of the box containing a cat that's dead and alive at the same time. You're feeling excited because you are about to open the box, your first quantum physics experiment. The possibility wave describing the state of your brain and the cat is given by:

$$|\Psi\rangle = (|🐈\rangle + |🐟\rangle)\,|😀\rangle$$

Let's also assume, that you happen to like cats. So what does Schrödinger's equation tell us about what happens to you after you open that box?"

Me: "The same thing that happened to Death, right? I become entangled with the cat."

Alice: "Bingo. If we don't arbitrarily kill possibilities, they stay alive and get entangled with you. The possibility wave turns into this:"

$$|\psi_{\text{🐈👤}}\rangle = |♟\text{👤}\rangle + |🐀\text{👤}\rangle$$

Me: "But now I am happy and sad at the same time because I saw a dead and alive cat at the same time. This is extremely confusing."

I picture myself traveling to Andromeda to have a support group meeting with Death and Shakespeare.

Alice: "On first sight, that's weird. And because this is not what we experience in everyday life, the popular Copenhagen interpretation just said that to solve this weirdness, all possibilities but one must magically disappear. Only one can manifest."

Me: "Which is equally weird."

Alice: "I agree. But we've forgotten something extremely important. We only considered you and the cat. But that's incomplete. We have to treat the universe as an unbroken whole."

I notice the waiter closing the door of the cafe and making a phone call.

Alice: "If we consider the rest of the universe, what happens is this. As soon as you open the box, the rest of the universe gets entangled with the cat as well. And that happens much faster than you can ever perceive.
Then the complete possibility wave is

$$|\Psi\rangle = |🧙‍♀️🙍‍♀️🌌\rangle + |🧙‍♀️🙍‍♀️🌌\rangle$$

Now you can see that inside of $|\Psi\rangle$, which describes all existence, we now have two different possible universes. One in which the cat is alive, and one in which it's dead. And as time goes by those universes become more and more different. In one of them you happily make your cat a lasagna, and in the other one, you get shot by Greenpeace."

Me: "Okay. But now there are two universes. That does not make any sense."

I wonder where Alice wants to go from there. Two universes? How will she get rid of the other one? Now she wants to go from killing cats and possibilities to killing universes? Sounds like fun.

Alice: "Well, what makes you so sure that two universes do not make any sense?"

Me: "Because there is only one universe."

Alice: "Are you sure?"

Is she asking me seriously if I am sure that there is only one universe?

Me: "Well, how can I be sure? There could be invisible universes all over the place. And there might also be an invisible flying Spaghetti monster just behind you. But doesn't adding them violate Occam's razor?"

Alice: "We are not *assuming* two universes. Schrödinger's equation *predicts* them. We are only assuming that the Schrödinger equation holds. The thing that violates Occam's razor is the additional assumption that some possibilities get killed. There is no equation predicting an invisible flying Spaghetti monster. At least no equation that's resisted a humongous amount of attacks, and for which thousands of predictions turned out to be correct. Tell me where I am wrong in my line of thinking."

She must be wrong. This is just too crazy.

Me: "So Schrödinger's equation predicts two universes if we manage to create a cat that's dead and alive at the same time. But that might be impossible to achieve."

Alice: "It also predicts two universes if we measure through which door a photon goes. Or if we measure any other superposition. Remember, we can already create those cat-states in the lab."

Me: "So you are saying that whenever a physicist creates a quantum state in the lab that contains two coexisting possibilities, the universe splits in two the moment somebody looks?"

Alice: "Not only then. Remember, quantum physics underpins all of physical reality. And humans are not special in being able to look. Everything can. Schrödinger's equation says, that whenever something gets entangled with something else, $|\Psi\rangle$ branches into more universes. And that happens unbelievably often every single nanosecond."

Me: "But what about gossip? I thought it turned 'ands' into 'ors'?"

Alice: "From our limited perspective it does. But in reality, gossip just drives the two universes apart. When you open the box, it's the spread of gossip that lets the whole universe find out about the cat. And once everyone knows, we can't go back, the two universes can't talk to each other anymore. They are completely different, in one everyone thinks the cat died and in the other, it's still alive. Gossip is the reason *it seems* that we are only living in a single universe, it means everyone in our universe agrees about what happened to the cat."

Me: "So the Schrödinger equation tells us that entanglement creates universes and gossip drives them apart."

Alice: "Correct."

Me: "But how can that be? That would mean that $|\Psi\rangle$ is not a possibility wave. It's describing a multiverse containing an incredible number of parallel universes."

Alice: "That's right. That's what Schrödinger's equation tells us. Everything not forbidden by the fundamental laws of nature is realized in some parallel universe. Of course, we haven't discovered every law yet, but the ones we have allow a lot."

Would they allow for my son to have survived the accident? For sure they would.

Alice: "You know what's really cool? There's even an app for the iPhone called the Universe Splitter. Everybody can download it. Do you sometimes find it difficult to make decisions?"

Alice pulls out her iPhone and shows me the screen. The app says it's able to split the universe in two. Before the split, you can tell the app what you are going to do in each universe. She enters: In one universe I will now: Kiss Bob. In the other one, I will now: Slap Bob[28].

Alice: "Let's say you need to make a hard decision. For example, accepting a job offer or not, or to fight with a crocodile or not. Or committing to a partner or not. With this app, you can do both at the same time. In different universes."

I do have difficulties making decisions sometimes. I know only too well how energy draining these internal battles can be. Meditation helps, but is there really a technological solution to this issue?

Me: "Crazy, how does that work?"

Alice: "The app connects to a laboratory in Switzerland. In that lab, there is a quantum device which sends a single photon onto a partially reflecting mirror. What happens is that the photon reflects from and passes through the mirror at the same time. But in different universes. Afterward, this device measures whether the photon reflected or not and tells you which universe you are in. Are you in the universe where the photon bounces off the mirror? Or are you in the universe in which the photon passes through?"

Me: "Okay, so this device splits the universe in two and then tells you in which universe you are in?"

---

[28] Nerd note: This app really exists.

Alice: "Exactly. And before splitting the universe, I committed that I will kiss you in the universe in which the photon passes through and I will slap you in the universe in which the photon reflects."

Me: "So if both versions of you are really committed to what you told the app, slapping and kissing will happen simultaneously? But in separate universes?"

Alice: "That's right. The device finds out in which universe we are in by measuring where the photon goes. This measurement is the first piece of gossip that drives the universes apart. Look, I will split the universe now."

Alice presses the big round button labeled "split universes". The app takes a couple of seconds. For a moment I wonder if I should kiss back. But that's the problem of another Bob in another universe. I can't react fast enough to dodge Alice's hand, and she gently slaps me on my cheek.

Me: "That was soft, Alice. I expected some more serious slapping. I'm a bit disappointed. I hope your kissing in the other universe is more sophisticated."

Alice: "Didn't want to annihilate you right away. I still might have a plan for you, you know? And I really hope you're not complaining about the kiss in the other universe. But I guess we'll never know."

Me: "This is completely nuts. You say that different universes don't talk to each other. But what about universes which are very similar?"

Alice: "Excellent question. Similar universes can talk to each other. But only when their differences are kept secret no gossip is allowed. Then the universes can even rejoin into a single one. Similar universes talking to each other and rejoining is exactly what brings about the weird quantum effects we can observe. The dance of possibilities behind the quantum curtain is, in reality, a dance of existing universes that are only different in a tiny way. This view beautifully explains the double-slit experiment. It explains how an electron can go through two doors at once because each possibility is just realized in a separate universe. The striped pattern we then observe is the result of two interacting universes joining together again. They can rejoin because the information about which door the electron goes through is kept secret from the rest of the universe. Instead, if any gossip gets out about which door the electron used, the universes can't rejoin, they become too different, and the striped pattern disappears."

I'm not sure if I can follow. My brain feels like a swamp.

Me: "What an insane map of reality."

Alice: "I agree. I find it hard to believe as well. Do you understand now why people attack the Schrödinger equation so much? And why people desperately try to modify it? And why people invented the classical stage even though we don't need it? But the fact is, to explain our everyday life experience, we don't need a mechanism which kills possibilities. There is just one everlasting unbreakable quantum wave $|\Psi\rangle$. Inside that wave there are multiple Bobs occupying multiple universes, having different experiences. Leading different lives. But the different universes can't communicate with each other. So each Bob thinks he is living in the only universe. And each Bob is convinced that he is the only Bob."

Does she really believe that? Is she sure that those hypothetical universes can't communicate with each other? I think about what Samantha said. *David loves you.*

Me: "Four shots please."

That might ease my overwhelmed mind. I wonder how many shots I ordered in the other universes. *In this one, Bob, you're an alcoholic.*

Alice: "It's crazy, isn't it? But that's what Schrödinger's equation tells us if we take it seriously. A single quantum wave describing all of existence, containing an incredibly large amount of different universes, which don't talk to each other anymore because of gossip."

Me: "But just because an equation predicts something doesn't mean it's true."

Alice: "That's right. But look at history. There are many things that the equations of physics predicted that people thought were crazy. Things that most scientists could not believe at the time, and sometimes for a very long time afterward. Maxwell's equations predicted electromagnetic waves[29]. Einstein's field equations predicted black holes[30]. The Dirac equation predicted antimatter[31], and Schrödinger's equation predicted entanglement. All of those things, even though they sounded absolutely crazy at that time, were found to really exist."

If this is true, my son would still be alive in many other universes. I wonder what David looks like grown up. Is there a universe where I hug him right now?

---

[29] Maxwell, James Clerk. "VIII. A dynamical theory of the electromagnetic field." *Philosophical transactions of the Royal Society of London* 155 (1865): 459-512.

[30] Schwarzschild, Karl. "Über das gravitationsfeld eines massenpunktes nach der einsteinschen theorie." *Sitzungsberichte der Königlich Preußischen Akademie der Wissenschaften (Berlin), 1916, Seite 189-196* (1916).

[31] Dirac, Paul Adrien Maurice. "The quantum theory of the electron." *Proceedings of the Royal Society of London. Series A, Containing Papers of a Mathematical and Physical Character* 117.778 (1928): 610-624.

Me: "So there really is a universe in which we are making out?"

Alice: "Yes and also one in which I would enjoy that."

Calm down, Bob. You don't know if your son is still alive. Nobody knows if there are multiple universes. But then it strikes me. As a baby, we think our family is the only family. Then, far in the past, we thought our tribe was the only tribe. Then we thought our village was the only village. Then we thought our continent was the only continent. Then we thought our planet was the only planet. Then we thought our solar system was the only solar system. Then we thought our galaxy was the only galaxy. And now we think our universe is the only universe. It would be entirely arbitrary if existence stops exactly here in blowing our damn minds. I notice that something starts to shift about the way I feel about myself.

Me: "But multiple Bobs sounds just too crazy. I mean what determines which Bob is me?"

Alice: "Nothing. There are many of you.

You are a multidimensional being living in many universes at once."

Me: "But I only experience one of them."

Alice: "What do you mean with 'I'?"

It hits me like an electric shock. What is the difference between me and an identical copy of me? My soul? But if quantum physics is true, then all the Bobs living in other universes would have the exact same soul. So am I really a multidimensional being? Is there really no difference between all the Bobs other than that their individual experiences at some point starts to deviate? And from then on, their brains just develop slightly differently. That renders the concept of an individual soul meaningless, Buddhism is right! But then reincarnation is not happening on a single timeline so that your body dies, and your soul enters another body. Not at all. Instead, it's much more powerful. It's the parallel existence of infinite beings having all possible experiences at the same time. You're never reborn. You always exist.

I take a deep breath. A part of me does not like this idea. It's the part which wants to be special. Which wants to be unique. It's the ego which feels threatened by the idea that it has no independent existence. The collection of memories creating the sense of "I am only this Bob". I know this part very well.

Me: "I mean the sense of 'I', which makes me feel that I am this person. And not anyone else. The feeling that this is me. That I am only this Bob and not any other Bob in another universe."

I am curious about Alice's view.

Alice: "I know this feeling of being a specific person. The feeling of identity. Everybody knows it. It is a powerful and deeply rooted feeling. But if the Schrödinger equation is describing reality, then it can't be true. Every Bob in each universe has this feeling of being the only real Bob. But none of them are. It must be a feeling based on a limited perspective. Based on incomplete information. Each brain is only perceiving a tiny slice of an unimaginable enormous multiverse. We only see the tip of the iceberg. And we can only identify with what we see."

Me: "Okay, let's assume for a moment that this is true. Is there a way to switch to another universe? I mean are those universes linked in any way?"

Switch to a universe in which my son is still alive? In which I feel at peace?

Alice: "You mean something like a universe hopper?"

Me: "Yeah."

For a fraction of a second, I notice Alice's eyes light up. Then she frowns. I wonder what she is thinking.

Alice: "What exactly would hop there?"

Me: "Me. I would just become one of the other Bobs. A Bob with a better life."

Alice: "You already are all the Bobs. Each Bob comes with his own brain. With his own memory. But those memories don't know about each other. Even if you were hopping between the universes all the time, you simply would not notice it. Because, as soon as you change universes, you also change brains. You change memories. And each memory tells you that you have always been in the specific universe it belongs to. The deeper question is: what is this 'you' that is supposed to hop and experience the perceptions and memories of each Bob?"

Very good question.

Me: "That can only be consciousness itself. We are consciousness."

Alice raises an eyebrow, clearly unsatisfied by my answer.

Alice: "That's hard to imagine. All those Bobs are already conscious. What would it mean for one consciousness to hop into another one? And what do you mean by 'we are consciousness'?"

How do you explain that to a scientist?

Me: "Well, you taught me earlier how the process of science creates a trustable belief system by rigorously attacking possible explanations and ideas."

Alice: "Right!"

Me: "In our brain, we don't just have a map of reality about the physical world. We also have a map of ourselves. Just like the map of the physical world, it consists of a complicated network of beliefs. Beliefs acquired through past experiences, the deepest ones from our childhood. So, the scientific way to approach the truth about ourselves would be to mercilessly attack those beliefs, right?"

Alice nods skeptically.

Alice: "Right. But how do we reliably disprove those beliefs? I mean you can't really do a physics experiment in some lab."

Me: "True. But let me ask you this: how do you know the results of a physics experiment?"

Alice: "You read them off some measurement device."

Me: "And the moment you read the results off a measurement device, they turn into human experience. That means that science relies on trusting human experience, right?"

Alice nods slowly: "That's right, I guess. Even if other scientists tell me that they see the same results, that would still be my human experience. In the end, that's all we have."

Me: "Good. What I'm saying is that your direct experience already is a reliable way to attack your map about yourself. The convenient thing is that you don't need to make the detour through the physical world. No labs required. The inconvenient thing is that you can't trust anyone else, only your own experience. Do you understand what I mean?"

Alice: "I think I do. For example, my direct experience clearly contradicts the belief that I am a gigantic pink elephant riding a monkey."

How does she know about my pink elephant fantasies?

Me: "So would you agree that if we have a reliable way of disproving beliefs, then we can create a belief system that's as trustable as science? Since this process uses the same principles, right?"

Alice: "That makes sense. Interesting."

Me: "Okay, then let's try that on the question of who we are. What do you believe you are?"

Alice takes a sip from her beer.

Alice: "Well. If I knew nothing about quantum physics, I would say I am this Alice."

Me: "What constitutes this Alice?"

Alice: "A body including a brain shaped by past experiences. Physically that's just a bunch of atoms."

Me: "Okay, then let's look at this closely. Let's start with your body. Focus on it. Do you experience yourself as one with your body?"

Alice closes her eyes. After a couple of seconds, she responds, "Well, not exactly. There is my body. But there is also something that's aware of my body."

Me: "Good. What is deeper? What is more fundamental? Your body, or the thing that's aware of your body?"

Alice: "Well, my brain creates consciousness, and it's part of my body. If my body dies, then my brain can't produce consciousness anymore, so that awareness of the body would disappear. So physically the body must be more fundamental."

Me: "Now you are back in your scientific belief system. You are making a detour through the physical world. Use direct experience. What are you *experiencing* to be deeper?"

Alice laughs, "I'm impressed. You are mercilessly attacking my beliefs. I love that."

Me: "Glad you like it. Most people would have killed me already."

Alice: "Not yet, Bob, not yet."

Why does she suddenly sound so serious?

Alice: "Okay, give me a second. I experience the awareness of my body as more fundamental because, without it, my experience of my body would be impossible."

Me: "Awesome. And the thing that's aware of your body is consciousness."

Alice looks intrigued, but at the same time, I sense fear. *What is she afraid of?*

Alice: "It kind of sounds reasonable."

Me: "Yeah! You can do this internal experiment with your thoughts, feelings, with your beliefs and any perceptions. With all the content of consciousness. There is always something that's aware of all those things. So you can't possibly be defined by any of those things, you must be beyond all of them."

Interesting. We switched roles. Nerds seem to be obsessed with the physical map of reality, while hippies focus on the map of reality we have about ourselves. That's where I am the expert. I love this conversation. Talking to someone like Alice really sharpens your mind.

Alice: "Okay. I get your line of reasoning. But that doesn't prove that you are consciousness, that it is fundamental.

It just suggests that that's a possibility. We can only *disprove* the ideas we have about ourselves. You could also be something that's even beyond consciousness."

What the hell is supposed to be beyond consciousness? If consciousness is the screen on which every perception appears, then beyond consciousness implies something which is watching the screen. Alice looks puzzled, but she seems excited.

Me: "That's true…"

Alice: "Anyway, consciousness or beyond, that's cool. I mean applying the scientific process via direct experience to the map of reality we have about ourselves. Thanks, Bob."

Me: "You're welcome. By the way, following this process leads to a technique called **self-enquiry**[32]. An old Indian tradition called Advaita. Check it out if you like."

Alice suddenly stands up, heading for the bathroom. For a second, I think about stealing her glasses.

It's funny how, in the end, I'm using the same process as Alice, just for a different goal. The purpose of science is to create a trustable belief system about the physical world.

---

[32] Maharshi, Sri Ramana. *Self Enquiry*.

The purpose of self-enquiry is to dissolve false and already existing beliefs about who we are. For me, this is proper spirituality without any esoteric bullshit. No woo-woo attached. In the end, both science and spirituality aim to improve the maps of reality we use to perceive the world. Even though both use the same fundamental principles in order to approach the truth, science is a constructive process while spirituality is a destructive process. Proper spirituality does not create a new map of reality, it just deconstructs an old one. It's the opposite of religious dogmatism.

Different brains carry different beliefs constituting different maps of realities. Beliefs range from 'I am a collection of atoms' to 'I am what I own', 'I am my job', 'I am my relationships' to 'I am worthless', 'I am not enough', 'I am ugly' to 'I am guilty', 'I am selfish' or 'I am evil'. Or, in the worst case, even 'I am a hippie' or 'I am a nerd'. Most brains carry most of those beliefs to a varying degree. Just like the gorilla who is trapped by his belief, those beliefs imprison us. They create a lot of unnecessary suffering. The deepest and most persistent belief is 'I am this body.' What would happen if we could dissolve all of those beliefs? What would remain? How would that feel?

What would happen if you were to successfully prove to yourself that these 'I am whatevers' can't be true? If only 'I am' remains without anything left to identify with? You would wake up from the dream of being something. You would just be. You would feel pure. You would not be limited anymore by any map. By any beliefs. That's what is meant by enlightenment.

The biggest obstacle in improving our maps of ourselves is that the beliefs we have are mostly unconscious, deeply carved out and emotionally charged. Pure intellectual understanding is not enough to dissolve these beliefs. If each belief is a physical circuit in our brains, then it is not enough to intellectually convince yourself that you are not your body or that you are not guilty. This just creates new circuits that compete with the old ones. The result is an enlightened ego. Someone who believes they've got it, but the deeply rooted false beliefs are still physically there, and life will sooner or later find a way to bring them back to the surface. The result: confusion, shame and cognitive dissonance. How do you solve this problem? How do you directly overwrite the old circuits?

I know I'm not anything close to enlightened. That thing I believe is me, is literally scared to die. I feel trapped in between pain and fear. The very real pain induced by false beliefs and the irrational fear of dying. *Your son died, Bob. You are guilty.*
I wave to the waiter. "A White Russian please." Less scary, but it works just as well. At least temporarily.

Has learning about quantum physics helped me at all so far? Coexisting possibilities? Cool! Entanglement? What the fuck! Parallel universes? Awesome! Infinite Bobs, living all sorts of different lives? Loving it. I had no idea that studying quantum physics radically changes your view of the world. That taking it seriously cures racism. Tames hooligans. Stops wars. I lean back and focus on my direct experience. I can see my body as just a perception. The person Bob is just a perception. Any thoughts and feelings are just perceptions. Even the sense of "I am this Bob" is just another perception. And then there is the awareness of all those things, consciousness. Simultaneously, there are sounds of cars and people. There is music playing, La Cafetera Roja. The spicy odor of Indian food. There is the table, pictures on the walls, the waiter talking into his phone in the corner. A body breathing. A deep feeling of relaxation. There is Alice, back from her bathroom vacation, shyly staring at her glasses on the table. I should have stolen them.

With the sense of "I am this person" unavoidably comes the feeling of separation. If I identify with anything, no matter if it's a person, a gender, a country, a football team or a race, then there is me and there is not-me. The 'me' can be threatened by the 'not-me'. It can be ridiculed, it can be judged, it can be destroyed. The sense of "I am this whatever" is responsible for feelings like fear, guilt, shame, hate and pride and ultimately for any conflict. Get over this shit man. World peace, motherfuckers.

Alice: "I hope you weren't bored."

Her eyes look as if she just smoked a joint. What did she do? Her vibes increasingly confuse me.

Me: "Don't worry, my mind doesn't allow boredom in my life. I was wondering, what's your view on consciousness? Where do you think it comes from?"

Alice: "I would say it's produced by complex biochemistry in the brain. But now I guess there really is an equally valid perspective, guided by direct human experience without making the detour through the physical world."

Me: "Right, the western world tends to see physical stuff as fundamental, the eastern world prefers consciousness. Like if you dream or if you play a video game. You seem to move inside a three-dimensional world. But the screen on which all the images appear is never moving. Only the content of the screen is changing giving rise to the illusion of motion. In the same way, everything, the galaxies, the stars, your body and even space and time could be inside of you, your consciousness."

Alice: "Well, that's certainly a valid view. But I don't agree if you say that the physical world is not real. That seems really unlikely to me. Why would an illusion follow some fixed mathematical laws? I strongly believe that there is something real about those laws. But I love the video game metaphor."

Me: "Sure you do, damn nerd."

Alice laughs: "You seem to know me pretty well already. So that's what eastern philosophy teaches us? Consciousness is fundamental? Interesting. Schrödinger shared that view as well."

> *"Consciousness cannot be accounted for in physical terms. For consciousness is absolutely fundamental. It cannot be accounted for in terms of anything else."*[33]
> **-Erwin Schrödinger**

Me: "Ah nice! This guy really was a genius. But yes, I'm convinced that's right."

Alice: "I'm not so sure. I prefer to admit that we don't know. Consciousness and matter could be lower dimensional shadows of something even bigger. Both perspectives could be true simultaneously. Just as a beer can's shadow

---

[33] Schrödinger, Erwin. *What Is Life? the physical aspect of the living cell and mind.* Cambridge: Cambridge University Press, 1944.

can be a rectangular and a circle at the same time, everything could behave like matter and consciousness at the same time. But who knows?"

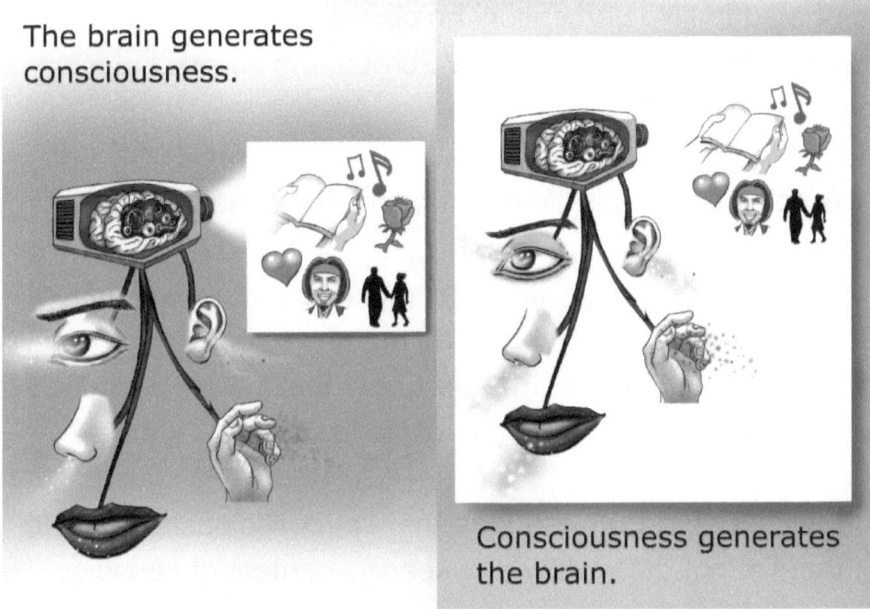

We don't know which version is true. Both perspectives could be lower dimensional shadows of something more fundamental, just like the shadows of the beer can.

Me: "Yeah, I suppose. Who knows."

Alice: "I like the way you are able to talk about consciousness, thanks for opening my mind a little. But science aside, I'm not sure I like the idea that 'I am only consciousness'. I mean, does it make sense to only identify with consciousness? Just being the screen without content? That feels a bit distant and detached from life, doesn't it?"

Me: "It does, but with this distance comes independence. Complete freedom of circumstances. That's the essence of wisdom. No perception, no circumstance can harm me. If a bomb explodes on a TV screen, the screen stays unaffected."

Alice: "Sure, but is life about being detached from it?"

It's true, identifying with nothing can become a way of hiding from life. During the three months I spent living in an ashram I felt peace. But did I only feel peace because I escaped from the struggles of life itself?

Me: "If I start from the idea that I'm consciousness itself, that I'm not defined by any specific content of consciousness. Then there are two equally valid perspectives I can take, two sides of the same coin. I am nothing. Or I am everything. The perspective that 'I am nothing' is wisdom. It sets you free because nothing can affect you. The perspective 'I am everything' is unconditional love. Because you are and you care about all of your direct human experience equally. And that makes you genuinely compassionate towards everything you perceive. But if you tend too much towards wisdom, you become cold and detached. You lack connection. And if you lean too much towards love, you lack depth, stability, and clarity. The Holy Grail is to see yourself as nothing and everything at the same time. Then you feel unshakable stability. Complete freedom of any experience. While at the same time feeling love and compassion for whatever appears inside of consciousness. This is the real Buddha. Wisdom soaked with unconditional love."

Alice: "So instead of saying 'I am an idiot', the correct way would be: 'The brain which seems to generate my experience right now, believes that it is an idiot. How adorable.'"

Me: "You got it."

Alice: "That's beautiful. Did you see the Yin and Yang on the window? *Be balanced or die.* I guess the average nerd is tending more towards wisdom."

Me: "And the average hippie towards love."

Alice: "We should spend the night together."

That was subtle. I almost missed it. *What is her plan? Creating a hippie-nerd?* Is that an experiment I want to be part of? I feel a sudden rush of heat flowing through my body. A wave of excitement? Or a warning signal trying to prevent me from making a horrible mistake?

Me: "Do you live here?"

Alice: "Yes, right around the corner."

Before I can object, she asks for the bill, pays a couple of thousand Satoshis, and we are out of the cafe.

Me: "Wait."

Alice: "What?"

Me: "What happens when we die?"

Alice stops and looks at me. She seems much more surprised by this question than I expected.

Alice: "I don't know. I think this is related to the question of how many consciousnesses exist. Does everybody have their own? Or is there only one? For me, it's hard to imagine that there is only one. Our bodies run around in this world in a way that makes it seem like they are independent, that each one comes with its own consciousness."

Me: "Maybe I can help you out there. Imagine a house with a lot of windows. Can you conclude from the number of windows, how many people are in the house?"

Alice: "No, of course not. But I don't get what you mean by that."

Me: "In the same way the number of windows doesn't tell you how many people are looking through them, you can't conclude from the number of bodies how many 'consciousnesses' look through them. In both cases, the answer might be just one."

Alice: "I get what you are saying. But that would mean that all people, all animals and plants and maybe even rocks and, of course, also all Bobs and Alices living in all the parallel universes are connected by consciousness. Not only connected, but that the same consciousness looks through all of them simultaneously."

> *"Consciousness is never experienced in the plural, only in the singular. Not only has none of us ever experienced more than one consciousness, but there is also no trace of circumstantial evidence of this ever happening anywhere in the world."*[34]
> **-Erwin Schrödinger**

Me: "Yeah, I believe it's possible. So what happens when we die?"

Alice: "Well, I guess in that case consciousness couldn't use dead bodies to experience the world in the same way

---
[34] Schrödinger, Erwin. "Mind and matter." (1958).

it can through a living body. The same goes for a body in deep sleep or an injured brain. Conscious experience in a body which is dead or in deep sleep stops. But globally, conscious experience goes on. And if you are really consciousness itself and there is only one consciousness as you say, your next possible experience must be an experience. You can't experience not experiencing. You would experience the process of dying, but then you would instantly have another experience. Maybe you'd find yourself looking through another configuration of atoms. A new body. But that new body only remembers being that body, because it comes with its own memory. And for you, it feels like you have always been with that body. So, consciousness would experience dying, but would never remember it."

Me: "Interesting. But could it be that that memory survives somehow? That all experiences leave a trace somewhere? Didn't you say earlier, that information never disappears and can only be passed on?"

I find it hard to interpret Alice's facial expression. Her cheeks are glowing, but she seems distant. It's frustrating to think that we could be bridged by a single consciousness, but I'm still no closer to knowing what she is feeling.

Alice: "That's true. Memory is information. Even the information once stored in the memory of a now decayed body is still physically somewhere. But it's totally scrambled up.

I have no idea if consciousness can still make sense of it. Or even have access to it."

While we are talking, we continue walking down Carrer de Paris. It's getting dark. I feel drunk. I pay no attention to the passing women. No attention to the traffic and no attention to anything else other than Alice's words.

Alice: "I would like to tell you a secret. But you need to promise not to tell anyone."

"Sure."

Alice: "I am working with a team on a device using quantum physics to do computations. A so-called **quantum computer**."

Me: "Ah, I've heard about these things. What's so special about them?"

Alice: "They're incredibly powerful. In a normal, 'classical' computer, information is encoded in bits. A bit is just a number, it can be either 0 *or* 1. In the quantum world, we can use quantum bits instead. They can take both values at the same time, they can be 0 *and* 1. The quantum computer we built is exponentially faster than all the classical computers on earth combined. Why? Because it's able to outsource computations to a vast number of parallel universes. And since our quantum computer is protected from gossip, those parallel universes keep talking to each other. That means that we can take advantage of all the results obtained in the other universes."

Me: "That sounds completely crazy. And that really works?"

Alice: "It really works. It can solve some problems in a second, for which the best classical computer needs more than ten billion years. It looks like quantum computers will revolutionize things like weather forecasting, traffic prediction and we can use them to find powerful new medicine by simulating chemistry. They might even find a cure for cancer or HIV. But many scientists still do not agree that a quantum computer uses parallel universes to get this incredible advantage."

Me: "Okay, so this parallel universes view is still not very popular?"

Alice: "It will be soon. Because just a month ago we achieved something amazing. We managed to run an artificial intelligence on our quantum computer[35]. Basically, we constructed a brain which is protected from gossip. A brain that can experience the splitting and rejoining of universes. We call it Qualin. We can ask Qualin questions. And it can answer us. It says it's conscious."

---

[35] **Beware:** Such a good quantum computer does not exist, yet. Thus, the following is science-fiction. However, it could very well be true.

Me: "So it can really experience parallel universes?"

Alice: "That's right. It experiences contradicting realities. Simultaneously. And it can talk about them."

What on earth is going on? Will this day ever stop blowing my mind? A conscious computer experiencing parallel universes? *What the hell does that mean?*

Me: "This is so damn crazy. Isn't that proof of parallel universes?"

Alice: "It seems like it. But it's much more than that. Remember $|\Psi\rangle$, describing *everything*. We can interpret the whole of existence as a giant quantum computer. And as a whole, it is protected from gossip, because there is nothing the whole can gossip to. Everything is on the dance floor, nothing can mess with its possibility wave. The whole is a perfect quantum computer."

Me: "Are you saying the whole of existence could be conscious like your Qualin? A consciousness that experiences everything? You, me and all the parallel universes simultaneously?"

Alice: "Exactly. And if we take your view on consciousness, it might really be, that you, me and everybody else, that we are all exactly that one consciousness."

We stop at an old building. I feel dizzy. I need some water. We take the elevator to the 6th floor, and we enter Alice's apartment. Her apartment is surprisingly cozy. The colors are warm, and there are even a couple of plants. The floor and the kitchen table are covered with papers. Taking a closer look, I see that they are full of cryptic mathematical symbols and strange drawings. There are $|\Psi\rangle$s everywhere. The universe on a piece of paper.

"That's what solving the Schrödinger equation looks like," Alice remarks, noticing my surprised face, "and those are my kickboxing gloves. So better behave."

She smirks. I get the sense she might also like the opportunity to test those gloves. But behave, I can do that for you, Alice. I can do that.

"I will do my best."

"Glass of cognac?"

"Why not."

"Wanna put on some music?"

"Sure."

I take out my dented mp3 player and connect it to her sound system. The rhythmic beats of downtempo psychedelic dub fill the room. We are drunk. We laugh. We balance wisdom with love in a mutual dance of contradicting possibilities. Suddenly, the world starts shaking. I sweat. I cough. There's something warm in my hands. Blood. More surprised than scared I look at Alice. Her relaxed smile has been replaced by stony frustration.

"Oh, that's unfortunate," she says, "I was really starting to like you."

My stomach feels like I swallowed a thousand needles.

"Unfortunate?! What is happening?" I ask.

"Unfortunately, Bob, that cognac you just had. It was poisoned. "

"What? Why?"

"Science."

"What the hell?"

"We want to know if and how consciousness is linking parallel universes." Alice looks at me.

"And so you poison me?"

"Don't worry. Before you drank the cognac, we split the universe into one where the cognac contains poison and one where it just contains our dear old friend ethanol. If it makes you feel better, it's like we cloned you and are now killing off one of the clones, so there's still the same number of Bobs."

It did not. My heartbeat is accelerating while everything else seems to slow down. The room gets brighter, and the music distorts into a cacophony of strange pitches and out of place beats. I begin to stumble.

"How did you split the universe?"

"You remember the Universe Splitter? That app for the iPhone? We connected a device that added poison to the cognac in the universe where the photon bounces off the mirror and left it untouched in the universe where the photon passes through. Now we know that we are in the universe where it added poison. Of course, we couldn't check beforehand which universe we were in. The placebo-controlled double-blind approach is the scientific standard.
This is a Schrödinger's cat experiment. With you as the cat."

"A fucking photon? Why me? Why didn't you try that on yourself?"

"What if I am wrong?"

Yeah! What if you are wrong??

I feel my body collapsing to the floor. Everything turns black. No white light. Nothing. Just nothing. Really nothing? There is still something which is aware of the nothing. There is still something thinking. There is still something noticing the thinking.

I wake up. I am alone in a king size bed. No blood, just a horrible hangover. It was only a dream. My head feels like it could explode at any second. I am tired, unbelievably tired. I get up and walk to the bathroom. The Barcelona water from the sink tastes heavily chlorinated, but I gulp it down all the same. Some cold water on my face feels good. I hear a noise and I smell coffee. I find Alice in the combined kitchen-living room making breakfast.

"Ah, you're awake finally. How did you sleep?"

"Not well. I had a terrible dream."

Alice stops what she was doing and focuses on me.

"What did you dream about?"

"I dreamed I was dying. You poisoned me."

"What did it feel like?"

"It was just black. It was like I was experiencing nothing. Pure emptiness."

"No, I mean the symptoms of the poison. What happened to your body?"

Why does she want to know that? I feel sick.

"Coughing, coughing up blood. Sweating. It felt like I had needles in my stomach. Then I collapsed and died."

"Interesting. Very interesting." Alice's eyes light up, and she seems to be trying not to smile as she turns back to the coffee.

What is going on? My headache is getting worse, that is for sure. I must lie down. I choose the couch, and while trying to make myself comfortable, I feel something hard below one of the cushions. A black folder. I want to put it on the floor, but something captures my attention. In its right corner, I see a small symbol. Suddenly, I am awake. Really awake. *How is that possible?* The Yin and Yang with skulls instead of dots. And right below in small letters, I read my own name.

"What the hell is that?"

"Nothing," Alice says sharply. She quickly tries to grasp the folder, but I am faster. I open it, and I see photos. Photos of me. My life. Alice moves for the door, but once again I'm faster, grabbing her shirt. This weak nerd has no chance. The punches come quickly, one fist then the other. I stumble backward, stars spinning across my view of Alice opening the door. *Oh right, the kickboxing.* Pain explodes from my nose, mixing with the nausea of the hangover. But then adrenaline takes over, giving me superpowers. I catch her from behind in the door frame. I force her arms behind her back and push her down onto her knees.

"What is going on?" I scream. "Why do you have all these photos of me?"

With the same symbol from the window of the cafe? The same symbol I saw on Samantha's hand?

I hear a noise behind me and quickly turn around. 'Speak of the devil' comes to mind, as I look directly down the barrel of a revolver. The arm holding the weapon belongs to a person I never expected to see again.

"Hi, Bob. You seem a bit confused," says Samantha. "I would suggest you calm down, sit on the couch, and then we can talk."

I let go of Alice and does what she tells me. After all, I don't have many options.

"Bob, we are very grateful for what you have done for us. You've helped us a lot," says Samantha.

"I don't understand." This is out of control.

Samantha: "Naturally. Let me explain to you, step by step, what is going on. By confirming the symptoms of the poison which killed you, you did a great service to science."

"Which killed me? But I'm alive!"

"Yes. You are alive. But you died in another universe."

"So it wasn't a dream?"

"No, Bob, it was a dream. But not a normal dream."

I need a drink.

Samantha: "Listen, Bob, you asked an important question last night. A question we've been thinking about for a long time. And today, you gave us a hint of the answer."

My brain just doesn't work under these conditions.

"Please just tell me what is happening."

Samantha: "Okay. The question we are obsessed with is this: does consciousness link parallel universes? If we could show that it's possible to experience a parallel universe in our dreams, that would be a big step forward. So, we turned you into Schrödinger's cat. One version of you died, and at the exact same time, the other version of you was blissfully asleep after drinking too much alcohol. If the version of you which survived dreamed about the version of you which died, then we would know that parallel universes are probably linked. And that we can experience them in our dreams."

Me: "And now since I've told you the symptoms of a poison I never received in this universe, you know your theory makes sense."

Samantha: "You got it, you're smarter than you look. But remember, you are only a single data point. Not worth more than an anecdote. The issue is far from settled. But it looks promising."

"I knew it. I saw your black soul right from the beginning. You fucking psychopath."

Samantha laughs.

"But why me?"

Samantha: "You're the perfect candidate. You have no family, no siblings, you lost your only son. You are a destroyed man. You're an alcoholic obsessed with sex, desperately searching for something you will never find. In the universe you died, nobody is going to miss you. We probably even did you a favor."

Even though that's all true, they are wrong. Feelings of guilt aside, I was actually quite happy most of the time. I guess humans can get used to anything. Whether you win the lottery or lose your family, after a while, you rebound to the same level of happiness.

"This is sick. How do you even know so much about me? I don't even have a smartphone!"

"That's actually perfect for us. Nobody is going to miss you on social media either."

"And your orgasm?"

"It was good acting, wasn't it? Giving you the impression that you can cause orgasms by looking and telling you that David loves you really got you. From your psychological profile, we knew you would respond to that. After all, you're obsessed with new age views. Accepting those ideas without criticality eases your pain a little, but it also made it very easy to hook you in. A symbol on the window of a cafe and a lonely, interesting looking woman was more than enough to make you walk right into our trap voluntarily."

I feel like shit. *Am I really that predictable?*

"But why did you tell me all those things about quantum physics?" I look at Alice.

"Error minimization," she begins timidly. "We thought it would be good for the experiment if you knew about quantum physics. We hoped that it might help in recognizing a dream about a parallel universe. After all, our data points rely on the direct human experience of someone else. We want to really understand what a dream caused by an experience in a parallel universe feels like. It also gave us a natural way to get you drunk so your brain would be in a receptive state. On top of that, I really enjoy talking about quantum physics. In fact, I loved our conversation. And to my surprise, I learned a lot from you."

What she says sounds honest. *But how insane is that?* They only taught me quantum physics to reduce the errors in their sick experiments?

"And now?" I ask, exhausted. Alice looks sad. I like her. I wonder what role she plays. Is she a victim of Samantha, too? I will never find out.

"We need more data points. And we can't allow you to destroy our project. But don't worry. You are not Bob. You are consciousness. You will be fine."

And with these words, Samantha shoots me right in my forehead. As soon as the bullet starts to penetrate the bone of my skull, time stops. I am flooded with emotions. I feel betrayed. Powerless. Angry. I want revenge. These monsters exploited my personality for their insane experiment, and after my job was done, they just shot me. I hate them with all the fibers of my body. It hurts. It hurts a lot. I can't bear it. I surrender. In slow motion, the bullet digs a tunnel of destruction into my brain.

Suddenly, I feel warm. Very warm. Almost cozy. I understand that everything that happens is just $|\Psi\rangle$ evolving. Choreographed by the sum of all energies. Feelings of hate and anger dissolve. After all, there are infinite parallel universes. And this Bob that is dying right now just happens to be in the same universe where Samantha and Alice turned out to develop psychopathic brains. Unbalanced brains. Too much wisdom. Not enough love. But after all, it's no one's fault. It's just $|\Psi\rangle$ evolving. I am the global eternal consciousness witnessing $|\Psi\rangle$ through $|\Psi\rangle$. I am looking through Samantha's eyes as I look through Alice's eyes. Dying feels like coming home.

Fundamental or not, there is no argument that consciousness is happening. Nameless, formless and identity-less consciousness which seems to perceive through the complex arrangements of atoms we call our body and our brain. And simultaneously space, time, our body and our brain are perceptions appearing on the screen of consciousness.

> *"We do not belong to this material world that science constructs for us. We are not in it; we are outside. We are only spectators. The reason why we believe that we are in it, that we belong to the picture, is that our bodies are in the picture. Our bodies belong to it. Not only my own body, but those of my friends, also of my dog and cat and horse, and of all the other people and animals. And this is my only means of communicating with them."*[36] **—Erwin Schrödinger**

We are walking down Carrer de Paris. We just arrived from a week of hiking in the Pyrenees. Life is good. Tomorrow I'll be back to teaching philosophy at the University of Barcelona. I love teaching. My primary interest is the philosophy of hedonism and the pursuit of happiness.

---

[36] Erwin Schrödinger. *'Nature and the Greeks' and 'Science and Humanism'*. Cambridge University Press, 1996.

"The search for happiness can make you crazy," I say. "In reality, it's very easy. There are awesome days, normal days and shitty days. That's it."

"You know what would make me happy, Dad?"

It's David's birthday. He just turned ten years old. I think about my 28th birthday five years ago. The day David almost died. I can't imagine my life without him.

"Ice cream?" I ask.

David smiles. He loves ice cream, especially in the heat of summer. I gently put my arm around his shoulder and bring his head to my side.

"Alright, let's get ice cream. Today is one of those awesome days."

For some reason I suddenly stop, finding myself in front of the window of a cafe. I notice this strange-looking young woman daydreaming about something she seems to enjoy. Following a spontaneous impulse, I wave at her. She waves back. Smiling.

THE END

Dear Esteemed Readers

Thank you, thank you, thank you! Without you, this book would be more pointless than a nerd without a hunchback. Therefore, thank you for your existence and your curiosity!

If you enjoyed this book

Are you in a universe where you know a hippie who has lost a bit of ground? Or do you know a spiritualist who is worryingly receptive to New-Age-charlatans? Or a nerd whose horizons could use expanding? Or do you simply think that world peace and humanity would benefit if this book were read as much as possible?

Then there are three things you can do:

- Recommend "Quantum Physics for Hippies"
- Gift "Quantum Physics for Hippies"
- Leave a maximum positive rating on Amazon (with a review if you like)

That would be incredibly kind of you, and we would be very grateful. Any of these actions would not only fill us with joy but also inspire us to write more books.

If you did not enjoy this book

If you find yourself in one of those dreadful universes where you did not enjoy the book: First of all, respect for your perseverance to get this far! You can vent your frustration directly to us at Lukas-neumeier@gmx.de. Please tell us about your experience with this book! We always take feedback seriously, and if it is sincere and constructive, we will include you in the acknowledgments of the next edition if you wish.

If you have questions, want to provide feedback, have suggestions, or want to insult us, you can also do so via the contact form at www.hippie-nerd.com. We read and respond to every single message.

Acknowledgments

Special thanks for extremely valuable feedback go to Kathrin Leitner, Sabrina Birk, Jonas Neumeier, Daniel Frank, Heinrich Langer, Anna Eisenhardt, Anton Dirnberger, Daniel Noske, Mark Müller, Christian Weisser, Barbara Chaplain, Jo Douglas, Susan Jacob, Thomas Rahn, and Christina Neugebauer.

Special thanks go to Klaus Groß. Klaus not only asked critical questions that ultimately improved our explanations but also helped us create www.hippie-nerd.de. He is the creative mind behind some of the products in our satirical online shop: www.hippie-nerd.de/shop.

More on Quantum Physics

"Quantum Physics for Hippies" offers a very deep exploration of quantum physics. Probably deeper than any other book on this subject - without losing itself in baseless speculation. Parallel universes may *sound* highly speculative, but they are indeed a prediction of the Schrödinger equation. Therefore, this book focuses on the most fundamental and important aspects and consequences of the Schrödinger equation.

Because it is important to me (Lukas) to illuminate quantum physics from multiple angles, I have decided to write another book. This time the focus is not on depth, but on breadth. The new book doesn't have a definite title yet, but it's more of a non-fiction book. It will contain many lively and understandable explanations of various quantum effects, including a section on quantum biology and current research. If you want to know more, just click on Dr. Lukas Neumeier on Amazon. There you can quickly see if the new book is already available. And if you like, you can leave us a short review on this occasion. Feel free to include an idea for a title! ;-). Peace!

www.ingramcontent.com/pod-product-compliance
Lightning Source LLC
Chambersburg PA
CBHW021817170526
45157CB00007B/2626